DISPOSABLE
BIOPROCESSING
SYSTEMS

DISPOSABLE
BIOPROCESSING
SYSTEMS

Sarfaraz K. Niazi

CRC Press
Taylor & Francis Group
Boca Raton London New York

CRC Press is an imprint of the
Taylor & Francis Group, an **informa** business

CRC Press
Taylor & Francis Group
6000 Broken Sound Parkway NW, Suite 300
Boca Raton, FL 33487-2742

First issued in paperback 2017

Version Date: 20111114

ISBN 13: 978-1-138-07700-3 (pbk)
ISBN 13: 978-1-4398-6670-2 (hbk)

Visit the Taylor & Francis Web site at
http://www.taylorandfrancis.com

and the CRC Press Web site at
http://www.crcpress.com

To Merlot Sinatra Niazi, the master of disposable emotions.

Contents

Disclaimer

While the author wishes to acknowledge the contributions of all of his peers, colleagues, and professional contemporaries whose works may have been quoted in this work, at times it is difficult to fulfill this responsibility, and the author is thankful to all those who have made this book possible. Included in this book are references to equipment used in bioprocessing; no guarantee is provided that the information is current and discussion of any particular piece of equipment does not constitute an endorsement.

Preface

Everyone's replaceable. Even you.

Unknown

Bioprocessing entails the use of a biologic entity to produce a target product as a by-product of the metabolic activity of the entity used. The science and the art of processing dates back thousands of years, from the fermentation of grapes by yeast to today's mass-scale production of monoclonal antibodies using Chinese hamster ovary cells. Recombinant engineering has made it possible to manufacture hundreds of life-saving endogenous proteins at a cost that is now affordable. However, the manufacturing of biological drugs (e.g., proteins and vaccines) is a difficult art to practice because the toxicity of these drugs is not always related to their chemical purity, but rather to the subtle variations in their structure, both three and four dimensional, that can produce serious immunologic reactions. Produced in recombinant cell lines and organisms, these proteins merely simulate, and do not always mimic, human proteins despite the use of the known genetic code to express these in host cells and organisms. A key concern of regulatory agencies, therefore, lies in assuring that there is no cross-contamination of the batches since it would not be possible to rely on any type of cleaning validation to assure that minute traces of substances would not affect the structure of the proteins. In most instances, we would not even know what the contaminants are.

The Food and Drug Administration (FDA) and European Medicine Agency (EMEA) thus strongly urge manufacturers to create environments that would keep the contaminants out rather than trying to clean them, and to show by validation protocols the effectiveness of the cleanliness. This stance of regulatory authorities became sterner in the 1970s as the issue of viral contamination came to the surface in the preparation of human- and animal-tissue-derived drugs. A large number of manufacturers who could not comply with the new requirements shut down, and a new awareness about the risks involved in the manufacturing of biological drugs arose. The companies that survived made huge investments in isolating manufacturing steps, continuous monitoring, and extensive viral clearance studies. The breakout of TSE further compounded the complexity and, as a result, it became extremely costly to manufacture biological drugs in facilities that would be BLA-compliant.

To assure compliance with the new regulatory requirements, major suppliers of components in drug manufacturing, like Pall, Sartorius, and Millipore, took the lead and developed disposable products that would eliminate the need to conduct cleaning validation exercises. The earliest products in this category were as simple as filters, and soon these became the standard components: today, more than 95% of filters used in bioprocessing are of the disposable type.

Before moving further into the historical perspective of disposable components, it is necessary that we review the regulatory definition of the term "single-use," which is only in context with devices. SEC. 201. [21 U.S.C. 321] Definitions states: (ll)(1) "The term 'single-use device' means a device that is intended for one use, or on a single patient during a single procedure." "Disposable" is defined by the *Oxford English Dictionary* as "made to be thrown away after use." Obviously, a single-use device is disposed of once its use comes to an end. A good example is a paper cup, which is disposed after it is used, but is there no reason why it could not be used a few times before it is thrown away. Similarly, how long can one reuse a disposable filter if the same buffer is sterilized by filtration over several days? The fact is that regulatory agencies do not require the use of single-use or disposable items in manufacturing; it is the responsibility of the manufacturer to assure compliance with limits of cross-contamination. It is when the cost and time required to meet those requirements becomes onerous that the cost of single-use or disposable items becomes a serious consideration.

Although over the past few years a greater number of components in bioprocessing are of a disposable type, these are still not in the mainstream of manufacturing for many reasons including the lingering questions about the quality of materials used, scalability, running costs, level of automation possible with these components, and the training of staff required to assimilate these components in an established bioprocessing system. The advantages are obvious: safer, greener, cheaper (particularly capital costs), and offering greater flexibility of operations. Perhaps the greatest impediment in the wider acceptance of disposable items comes from the inability of manufacturers to discard their large investments made, relatively recently (1970s and 1980s), in fixed equipment and systems. As a result, the changes that are taking place are at the level of smaller companies, research organizations, and contract companies. However, this is about to change rapidly. The high cost of production that was acceptable to Big Pharma must now be challenged as the patents of blockbuster recombinant drugs have begun to expire, allowing smaller companies to compete on cost with Big Pharma. The generic business in biological drugs should convince Big Pharma to adopt what I predict to be the future of bioprocessing. There are also environmental considerations involved. For example, Amgen's facility manufacturing Etanercept in Rhode Island consumes 800,000 gallons of water per day, most of which is to perform sterilization-in-place (SIP)/cleaning-in-place (CIP) and operate autoclaves. None of these would be needed in the new generation of disposable systems.

This book is the first attempt to consolidate the state of the disposable bioprocessing industry, to make the reader aware of the controversies, misconceptions, costs (capital and running), regulatory considerations, and the choices available now and those coming in the future.

The author has had firsthand experience in establishing the first max-disposable manufacturing facility for recombinant proteins in the United States. The "max-disposable" is another aspect related to choices to be made, keeping in mind that the purpose of switching to disposable technology is to reduce the cost. Whether it comes from a lowered regulatory barrier, capital investment, or running cost is irrelevant. Overkill in using disposable items would not be advisable, and this advice is provided throughout the book. Dispersed throughout the book are descriptions of the innovations introduced by the author to the bioprocess industry that range from the world's first stationary 2D bioreactor to preparative bioreactors to novel manufacturing layouts; the reader may read about these innovations at the U.S. Patent Office database or write to the author without any obligation.

This book is arranged in a manner to a newcomer ready to adopt disposable systems, every piece of information and knowledge in making good judgments.

Chapter 1. The Bioprocessing Industry—An Introduction. The current state of the use of disposable systems is described to bring the reader immediately to a level of understanding how others are doing it; also provided in this chapter are the resources available to readers to further their knowledge.

Chapter 2. Safety of Disposable Systems. It is important to understand what constitutes the greatest challenge in adopting disposable systems; this chapter deals in detail with the problems associated with the use of plastics or elastomer systems; the facts, the myths, and the road to assuring regulatory compliance are provided here.

Chapter 3. Containers. Disposable systems are most widely found in containers used in routine processes from mixing of culture media, buffer, and refolding proteins to storage of in-process and finished product. Since the container must be compatible with the product, these components require careful selection. This chapter describes various uses, the advantages of using disposable containers, and suggests several novel uses of disposable containers in biological processing.

Chapter 4. Mixing Systems. Advantages of using a mechanical device that need not be sterilized and reused made the development of several novel devices to mix the contents in disposable bags; the choices range from impellers to magnetically levitating spinners and air flow mixers. This chapter describes the relative advantages of each of these systems with intent to make the process components as cost-effective as possible.

Chapter 5. Disposable Bioreactors. The most significant impact in bioprocessing comes from using disposable bioreactors; still in their infancy stages because of the limitation in size, integrity, and safety considerations, this is going to be the most significant component of future bioprocessing needs. This chapter describes a brief history of bioreactor development and discusses the reasons for choosing the two-dimensional flexible bags as the true game changer of the industry. Provided in this chapter are the details of all current offerings and a guide for choosing bioreactors.

Chapter 6. Connectors and Transfers. Devices used to transfer materials from one vessel to another such as tubing, connecters, tube sealers, etc., play a significant role in designing a complete disposable bioprocessing chain. Since these components have been around for the longest time and their utility well established, it is easier to choose correct components since these are also subject to the same safety evaluation as the bioreactors. This chapter describes relative merit of different materials used in the manufacture of these components and advises on making an appropriate choice.

Chapter 7. Controls. Controlling processes in a disposable system offers many challenges because parts of the control systems also need to be disposable; this is an emerging field of invention and the users are likely to see substantial advances in the near future.

Chapter 8. Downstream Processing. While most advances in disposable bioprocessing have occurred in upstream processing, only recently have we begun to see choices made available for downstream bioprocessing as well; from disposable columns and media to skid components, a variety of these choices are now available. This chapter advises on deciding whether it is appropriate to consider disposable downstream systems because of the high cost and diminishing returns on the efficiency of these systems.

Chapter 9. Filling and Finishing Systems. The manufacturing systems for bioprocessing fall under the purview of equipment suppliers who are generally not the suppliers of the systems used in converting biological raw materials into products ready for use in humans; there is major gap in the art available for disposable manufacturing of biological drugs. Several new offerings, some made available only very recently, now make it possible to reduce one more regulatory barrier in the manufacturing of biological products. New products in this field are introduced in the book.

Chapter 10. Filtration. One of the earliest devices that went disposable was the filter since it was difficult to clean and re-use; however, with expanding choices of filters for culture media, buffer, and the finished products, it is important to know how to choose a compatible system that will provide the most cost-effective solution. This chapter provides selection criteria and suggests many options for different types of products.

Chapter 11. Regulatory Compliance. The largest cost-savings in the use of disposable systems comes from reduced regulatory barriers; generally not accounted for in the overall design of bioprocessing systems, this aspect requires a greater understanding. This chapter describes how using disposable systems will allow companies to expedite drug development, reduce turnaround time, and provide a cost-effective solution to small- and large-scale manufacturing of biological drugs.

Chapter 12. Environmental Concerns. Blown out of proportion, the environmental concerns in the use of disposable bioprocessing components is minimal given the overall use of other disposable items, from plastic bags to

bottles to paper products. The concerns about disposition of plastic components and their biodegradation are discussed in this chapter to alleviate any
moral concerns in the use of disposable bioprocessing systems.

Chapter 13. The Epilogue. A recap of theme presented in the book is provided here with predictions for the future of bioprocessing industry and
predictions that in the future biological drugs will be produced using only
disposable systems; advise is given to both large pharmaceutical companies
and small developers to begin planning a switch to disposable systems as
early in their plans as possible.

I am highly grateful to T. Michael Slaughter of CRC Press for encouraging
me to write this book and giving me this remarkable opportunity to share
a lifetime of experience with my readers. This book is a practical manual
that will be found just as useful as a handbook as it would fit in a teaching
curriculum.

The great team of editors at CRC Press always makes great contribution to
the final published form; Laurie Schlags, Kathryn Younce, Susan Horwitz,
and others who made significant contribution to this book are greatly
appreciated.

The information contained in this book on the disposable component is
derived from the data provided generously by GE Healthcare, Pall, Sartorius-
Stedim, Millipore, and many others; the reader is advised to always consult
with their websites regarding any changes to specifications and also regarding availability as all of these companies are fast changing their portfolio
of products. By mentioning commercial equipment as an example, I do not
intend to endorse these products and equivalent products by any reputable
manufacturer would perform as well.

I would remiss if I did not acknowledge the support of the great scientists
and leaders at Therapeutic Proteins Inc., the first max-disposable company
located in Chicago and utilizing over a dozen "game changing" inventions;
I would like to thank my team of scientists (in alphabetical order) Aleksey,
Ali, Brian, Carl, Daniel, Erum, Irwin, Jason, Miadeh, Naila, Nadia, Nicole,
Omayr, Paul, Rachel, Raj, Ron, Rosa, Stutee, Sunitha, Thomas, and Zafeer,
and the folks at Therapeutic Proteins Inc., for their assistance in helping me
develop the innovations and inventions described in this book and generally
allowing me to validate many suggestions that I have made in this book. The
support and guidance provided by Steve and Daniel Einhorn, Teresa Essar,
and Alvin Vitangcol are highly appreciated. Thanks are also due to Kevin
Ott and other members of BPSA (BioProcess System Alliance). The assistance
of Omayr Niazi in proofing the book, as always, was invaluable.

This book can be considered a sequel to my book *Handbook of Biogeneric
Therapeutic Proteins—Manufacturing, Testing, Regulatory, and Patent Issues* that
was also published by the CRC Press and found a large audience in small
and large pharma and biotechnology companies, regulatory agencies, teaching institutions, and contract organizations. I hope that my readers will find
this book just as informative and useful.

While I have taken care to make the information provided as current and correct as possible, mistakes would inevitably occur; I shall be highly grateful if readers would bring these to my attention by sending me an e-mail to niazi@pharmsci.com.

I have dedicated this book to Vijay Singh, the inventor of Wave Bioreactor, who literally showed the industry how to think outside the box—by removing the stainless steel vessel; by adopting a 2D flexible bag to work as a bioreactor, Vijay Singh removed the box around the materials essential to upstream processing. Feel his presence in the scores of inventions that I have made adding many new functions to his 2D flex bag.

Sarfaraz K. Niazi, Ph.D.
Deerfield, Illinois
May 10, 2011

Author

Sarfaraz K. Niazi has been teaching pharmaceutical sciences and conducting research in the field of drug and dosage form development for over 35 years. A former professor at the University of Illinois, Niazi has written over a hundred papers, dozens of books, and owns dozens of patents for his inventions in the field of drug development and biopharmaceutical processing, including patents on novel bioreactors. His first book on the subject, *Handbook of Biogeneric Therapeutic Proteins* (CRC Press), was widely received as a primer in the field of biological manufacturing. Niazi has hands-on experience in designing, establishing, and validating biological manufacturing facilities worldwide. He lives in Deerfield, Illinois.

1

The Bioprocessing Industry—An Introduction

A soul is but the last bubble of a long fermentation in the world.

George Santayana

The discovery of the DNA structure in the middle of the 20th century led to numerous breakthroughs in biological science and inspired a generation of entrepreneurs. The 1980s and 1990s saw a booming biotech industry introducing many biologic products to the market. As with small-molecule drugs, biologic development faces challenges in long development cycles, low success rates, and high costs of development that clearly surpass the billion dollar mark. Despite this financial barrier, the biological drugs industry continues to thrive; it is anticipated that in the future almost 40% of all new applications would be for biological drugs.

The 2010 sales of mainly recombinant therapeutic proteins and antibodies exceeded US$100 B (from $92 billion in 2009 to $108 billion in 2010). Growth was mainly driven by therapeutic antibodies (+16% to +33% versus the previous year), which accounted for 48% of biologics sales in 2010. Among the therapeutic proteins, double-digit growth was reported for insulin and insulin analogs (+17%) and recombinant coagulation factors (+16%), whereas modest growth (4% to 7%) was observed for therapeutic proteins, except for erythropoietin, which continued its descent (–3% versus 2009) and follicle stimulating hormone (FSH) products (–1%). The anti-TNF biologic etanercept continued to be the single best-selling blockbuster molecule with 2010 sales of US$7.287 B. The insulin analog detemir achieved for the first-time blockbuster status, and increased, together with the neurotoxin Botox, the number of blockbuster antibodies and proteins to 30. Such spectacular growth of biological drugs also comes with a forecast that in the future more than 40% of all drugs approved would be derived from biological sources.

The engine for biological manufacturing comes from ever-improving expression systems, and Table 1.1 gives examples and their status as of today. While the barriers to developing new drugs keep getting higher because of the regulatory demands of assuring safety, the technological barriers to manufacturing these drugs have certainly come down. The current technology can be traced back to the dawn of civilization, through mammalian cell culture technology—the expression system preferred for most known therapeutic proteins with desirable glycosylation patterns—is relatively new. It took two decades of trials and tribulations to bring cell culture from a bench

1

TABLE 1.1

Recombinant Production Engines

Host Organism	Most Common Applications	Advantages	Potential Challenges
Cell-free	Rapid expression screening; toxic proteins; incorporation of unnatural labels or amino acids; functional assays; protein interactions	Rapid expression directly from plasmid; open system: easily add components to enhance solubility or functionality; simple format; scalable	Expression yields over 3 mg
Bacteria	Structural analysis; antibody generation; functional assays; protein interactions	Scalable; low cost; simple culture conditions	Protein solubility; minimal posttranslational modifications; may be difficult to express functional mammalian proteins
Yeast	Structural analysis; antibody generation; functional assays; protein interactions	Eukaryotic protein processing; scalable up to fermentation (g/L); simple media requirements	Fermentation required for very high yield; growth conditions may require optimization
Insect	Functional assays; structural analysis; antibody generation	Posttranslational modifications similar to mammalian systems; greater yield than mammalian systems	More demanding culture conditions
Mammalian	Functional assays; protein interactions; antibody generation	Highest level of correct posttranslational modifications; highest probability of obtaining fully functional human proteins	Multi-mg/L yields only possible in suspension culture; more demanding culture conditions

technique at milligram scales to industrial production at kilogram scales. The era of biopharmaceuticals is manifested in the capability of producing large quantities of biologics in stainless steel bioreactors. Today, those large-scale stirred-tank bioreactors (usually >10,000 L in scale) represent modern mammalian cell culture technology, a major workhorse of the biopharmaceutical

industry. Many blockbuster biologics—such as Enbrel (etanercept from Immunex Corporation), Avastin (bevacizumab from Genentech (Roche)), and Humira (adalimumab from Abbott Laboratories)—are produced using large-scale bioreactors. The current state of manufacturing thus represents the peak of what we conveniently call the "age of stainless steel."

The method of manufacture of biological drugs progressed through an expected route. Fermentation in large vats, whether it was done for wine or industrial chemicals or drugs such as penicillin, was a well-established technique, so when the time came to manufacture recombinant drugs, the same systems were transported over to this new class of drugs around 30 years ago. Large stainless steel fermenters were a good fit as their science and technology was well developed. However, lurking in the bush was a new enquiry by major regulatory agencies: the quest to control cross-contamination and viral clearance, the two most important causes of the side effects of these drugs. The quality guidelines by the FDA and EMEA began emphasizing the safety issues for cleaning validation and viral clearance, and the industry responded with more robust validation plans to prove compliance. The costs of manufacturing soared, but that did not make any difference because all of these molecules were under patents, and the companies were able to get whatever price they needed to justify these huge investments.

However, the honeymoon for the biological manufacturing industry began to end with the expiry of patents and the eagerness of the EMEA to start awarding generic approvals of these drugs; suddenly, the cost of production did become a consideration.

While the stainless steel manufacturers reaped huge profits selling their multistory fermenters and bioreactors, the industry of flex-bag drug formulation and administration and of intravenous bags also thrived. However, few saw the need to connect the two, for there was no financial incentive to do so.

The first "disruptive" innovation came to the industry when the first disposable Wave Bioreactor™ was introduced in 1996, which coincided with the highest ever number of biotechnology drugs approved in a single year between 1982 and 2007. Almost immediately, the biological manufacturing industry (and more particularly the stainless steel industry) began a debate on the safety and utility of plastic bags to manufacture biological drugs, and the greatest fear inculcated in the heart of prospective users was the issue of extractables and leachables, a topic that gets a detailed review in this book. Ironically, this issue was long resolved, when the FDA allowed the use of plastic bags to administer drugs of all types, of both aqueous and lipid origin and including hyperalimentation solutions. The risks to patients were minimal vis-à-vis the convenience of administration. In reality, the leachables in biological manufacturing are of little importance as the exposure to these possible chemicals comes at a very early stage in the production, and the robust downstream purification that removes even the isomers of the compounds is more than adequate to remove these contaminants. The greater risk lies in the interactions in the final dosage forms. A notable incidence was the reporting of pure red cell aplasia (PRCA) in using erythropoietin and, while many causes were brought to attention, one was the interaction between the rubber stopper and the newly formulated drug containing a new surfactant that might have extracted some extractables from the rubber stopper.

The fear of leachables, the strong presence of a well-established stainless steel industry, and a user industry in no rush to learn how to reduce the cost of production slowed down the implementation of plastic containers, more particularly of the disposable containers in drug manufacturing.

First came the changes in practice as the industry began using disposable filters, flexible containers, membranes, sampling devices, and now there has been a wave of disposable bioreactors to address the most critical barriers in biological drugs manufacturing. The stainless steel industry remains robust today, thanks to the reluctance or perhaps the inability of Big Pharma to junk their dinosaurs and give in to the "disruptive" technology for upstream manufacturing that first appeared as the famous Wave bioreactor that utilized a rocking platform and a 2D flexible bag in 1998. (In this way, the industry owes much to Vijay Singh, the inventor of Wave technology.)

Disposable bioreactors have since evolved beyond the wave-based design and have been adopted both for research purposes and Good Manufacturing Procedures (GMP) production. Other disposable technologies, such as disposable filters, flexible containers, membranes, sampling devices, and chromatography columns, have also made a significant headway in being accepted as the standard of manufacturing.

The final decade of the 20th century was good for the biotechnology industry, which raised billions in the public market, and a rush for new regulatory filings was soon on; however, many of these companies did not have in-house expertise to manufacture these molecules and that caused a mushrooming of contract research organizations (CROs) and contract manufacturing organizations (CMOs) that were ready to fill the gap. It became relatively easy to secure clinical test supplies without having to construct a recombinant manufacturing facility: this eased the financial pressure as well as made up for the dearth of qualified individuals in this newly found science of manufacturing. However, CROs and CMOs could not afford the capacity of large stainless steel technology since they would not know which product they would be handling the next day: disposable became very popular (because they required so little capital investment) among the CRO/CMO groups as well as research organizations, even though their need for regulatory compliance was less.

The improved efficiency of being able to switch over to different products and manufacturing methods pushed the equipment supplier industry to make some quick innovations. The list of disposable items expanded very quickly, and we can readily classify them in three categories.

Category I includes well-established disposables that came a long time ago, and these include analyzer sample caps, culture containers, flasks, titer plates, petri dishes, pipette and dispensing tips, protective clothing, gloves, syringes, test tubes, and vent and liquid filters.

Category II includes line items that were necessitated by the problems in cleaning validation. These became fully accepted about a decade ago and included aseptic transfer systems, bags, manifold systems, connectors, tri-clamps, flexible tubing, liquid containment bags, stoppers, tank liners, and valves.

Category III includes the most recent trends within the past five years and includes bioprocess containers (though the first one was introduced in 1996 by Wave, it became mainstream only after GE Healthcare bought Wave), bioreactors, centrifuges, chromatography systems, depth filters and systems, isolators, membrane adsorbes, diafiltration devices, mixing systems, and pumps.

There are many published surveys of the industry reported in the literature and, while these statistics can be tainted because the equipment suppliers support most of these, a few general trends that are established include [please refer to the Bibliography on the sources of these surveys.]:

This is the current state of the use of disposable systems as of 2010 (BioProcess International Survey summary):

1. The use of disposable bioprocessing is growing at the rate of 30% per year.
2. The biopharmaceutical or biological manufacturing industry consumes almost one-third of all disposable products used, followed by the biodiagnostic industry.
3. The CROs are least likely to use disposables because of the capital cost investment and the fact that they are used to the adaptability of the hard-walled systems.
4. Most of the adaptations of disposable technology are in the United States, comprising two-thirds of all worldwide use, and with Europe a distant 50% of the United States.
5. The companies with fewer than 100 employees constitute about one-third of customers and so are the companies with more than 5,000 employees; the midsize companies are taking longer to evaluate the merits of disposable systems.
6. Three-fourths of the companies using disposable systems are using these for manufacturing, with less than 10% of companies involved in drug discovery using disposable systems.
7. Companies with more than six products account for almost 60% of all disposables used.

8. More than 80% of new products utilize disposable systems, and almost 70% of existing manufacturing processes have been modified to include disposable systems.

9. The main concerns about the use of disposable systems in the order of importance are
   a. Capital investment
   b. Experience in using these systems
   c. Validation and environmental concerns
   d. Concern about leachables and extractables
   e. Integrity of systems

10. European regulatory agencies, as well as European companies, have greater concerns for leachables and extractables, and while the FDA allows greater flexibility in adopting newer systems, the EMEA has drifted away from common acceptance criteria.

11. The most widely used disposable components are bags and bioprocess containers, followed by filters (which constitute the main cost concern), connectors, bioreactors, mixing vessels, chromatography, and sensors. This trend shows that the simplest of the components, which require little problems in validation, are the easiest to adopt; obviously, chromatography and the use of disposable sensors would present a much high barrier to validation.

12. The unmet needs of the industry in adopting disposable processes include
   a. Leachables
   b. GMP compliance of disposable sensors, calibration scale
   c. Robustness of sensors and chromatography equipment
   d. Reliable bioprocessors that are cheaper
   e. Scalability
   f. High volume and flow rates
   g. Lack of single-pressure flow and temp transmitter
   h. Larger scale, greater than 100 L
   i. Standardization
   j. Lab scale, less than 3 L

13. Most companies have allocated less than US$100 K for disposable products.

14. The main reasons for adopting disposable systems:
   a. Cleaning/sterilization cycle
   b. Convenience
   c. Flexibility

d. Operating cost

e. Capital cost

f. Turnaround time

g. Reduced process steps

h. Smaller foot print

i. Rapid scale-up

j. Improved environment

15. Selection of specific disposable systems depends on

a. Availability

b. Price

c. Quality

d. Approved supplier

e. Documentation

f. Customer service

g. Product offering

h. Past purchase history

i. Engineering support

16. The disposable systems are disposed of 57% by incineration, 37% by landfill, 20% waste to energy, and 10% converted for alternate purposes.

17. Over 80% of users are satisfied with their adoption of disposable systems and have demonstrated savings in cost.

18. The main misconceptions in the use of disposable systems include

a. May be more costly over time, specially filtration

b. Not sure of savings

c. New investment needed

d. Have no need to save cost

19. The main regulatory concerns about disposables include

a. Sterilization and extractables/leachables

b. Leaking of containers

c. Bag integrity

d. Validation of sterility/manufacturing process

e. Aseptic process validation

f. Quality of multiple suppliers

g. Validation, lot-to-lot variability

h. Material compatibility

i. Reproducibility of batch process

The main driving force behind the growth of disposable systems remains the cost benefit, despite a myriad of analyses presented in the literature with remarkable shifts to which point of view the author subscribes to or the industry sponsoring the publication. There is no dearth of what we have come to know as "advertorials" describing new technology and its benefits. Some companies have done better than others. A Google search for "Cultibag" and "Wave Cellbag" returns about the same hits, around 6,000; a search for "Xcellerex" provides about 60,000 hits. While the marketing themes may differ considerably among companies, there is no doubt that some have touted their products too much and demanded unreasonable prices when equally robust and much cheaper alternates are available. One of the purposes of this book is to point out to readers those differences.

Next to the total cost, which is significantly lower, is the attraction of timeliness in the use of disposable components. Ready and available components that require little preparation make it easier to switch over applications easily. The newest concept is to offer a complete line of solutions as offered by all of the major suppliers (GE Healthcare, Sartorius-Stedim, Pall, and Millipore). Before examining the strength of this streamlined system, it would be educational to examine how Big Pharma currently plans its manufacturing systems.

A case in point is Amgen's retrofitted recombinant protein manufacturing in West Greenwich, Rhode Island. Amgen invested about US$500 M to enable manufacture of its blockbuster drug Enbrel (etanercept), whose patent expires worldwide in 2012. Amgen's plant is now one of the biggest mammalian protein manufacturing plants in the world. The project involved retrofitting an existing facility (the BioNow project) and the construction of an entirely new manufacturing plant as well (the BioNext project). Both plants use Immunex's T1 Enhanced Process developed by the Immunex Process Science Group. Amgen uses 800,000 gallons of water per day in this facility and that has brought about a conflict with Kent County Water Authority as an example of how these megaprojects affect both the environment and the cost of production. The current facility includes a production building, a warehouse, a central utility plant (47,000 ft^2), and a quality laboratory. The area covered is 500,000 ft^2. Also included are nine bioreactors with a capacity of 20,000 L each; this is about ten times the size of bioreactors currently used in most pharmaceutical manufacturing plants. The Kinetics Modular Systems provided smaller bioreactors—3,000 L and 15,000 L—and the harvest module. The retrofit involved the adoption of 120 pieces of major equipment, 25 mi of pipe, 240 mi of electrical wire, and 300 tons of heating and cooling ducts.

The Wyeth biotechnology campus opened in September 2005 and is currently the largest in Europe. The campus, which makes Wyeth the largest pharmaceutical employer in the Republic of Ireland with 1,370 employees, comprises a development facility as well as a drug substance and drug production facility, representing 1.2 million ft^2 in building space. The site is the largest dedicated biopharmaceutical development and manufacturing investment in the world.

Wyeth invested US$1.8 B in its Grange Castle facility, where site development work began in October 2002. The campus comprises three separate facilities: a drug development unit, a drug substance unit, and a drug production facility. These facilities went into production on a phased basis by 2009. Products that are manufactured at the new biotech facility include Enbrel (etanercept), Prevenar (pneumococcal conjugate vaccine), antihemophilic factor VIII, recombinant human bone morphogenetic protein (rhBMP-2), Tygacil (tigecycline IV), and Relistor (methylnaltrexone bromide). Construction required more than 15,000 tons of structural steel, 160,000 ft of process piping, 2,400 items of equipment, 7,587 engineering drawings, and 1,200 specific validation protocols. Zenith Technology was responsible for the validation of all automated systems. This new facility is the largest fully integrated facility ever built in a single phase. In March 2007, the Wyeth Corporation announced a further $32 million investment at the Grange Castle site, which would include the construction of an additional 6,000 m^2 of R&D laboratory space. This will then take the total laboratory space available at the site to 8,500 m^2.

The examples given earlier show the complexity of recombinant manufacturing projects. The high cost of these facilities is well reflected in the price of these drugs in the market and, as long as new molecules keep getting approved, the trend to construct bigger and bigger facilities would continue.

However, a new phenomenon is happening in the industry with the rise of biogeneric or biosimilar drugs that would inevitably put pressure on Big Pharma and, unless they adopt a more cost-effective method of manufacturing, they will be priced out of the markets. Unfortunately, the huge infrastructure that Big Pharma needs to support would make it difficult to cut the cost down and, as a result, there is a new type of partnership developing whereby Big Pharma would outsource the manufacturing of its existing or even new Active Pharmaceutical Ingredients (APIs) to smaller companies who are better prepared to adopt the newer technology such as the use of disposable systems.

One way to look at the value of disposable system is to make comparisons of the time it takes to complete a batch. For example, it takes about 50 hours for a batch of a monoclonal antibody to reach from the end of the upstream stage to purification; using disposable systems, this time can be reduced by at least 50% by using the GE ReadyToProcess system, an integrated system of disposable components. Similar systems are offered by Sartorius-Stedim, Pall, and EMD Millipore, the details of which are provided in Appendix I.

Another advantage of disposable technologies is their portability. The floor plan of a disposables-based facility can be changed much more easily than that of a traditional facility. Different process requirements can easily be addressed by moving equipment into or out of a production suite. Because of the disposable nature of disposable systems, contamination is less of a concern (especially cross-contamination for multihost, multiproduct facilities).

Disposable systems were first accepted by process development and production groups for toxicology studies and early stage clinical trials. As commercially available systems become more robust and reliable, disposables have been incorporated into process platforms by many manufacturers, and more commercial production facilities now use these technologies as an integral part of their manufacturing processes and of their efficiency and productivity improvement tools.

The most significant change in the disposables environment has come in the form of disposable bioreactors. With lower capital investment, ease of operations, and portability, they are likely to replace stainless steel stirred-tank bioreactors. Flexible containers will come with presterilized assemblies such as ports, filters, and sensors for storage of buffer and product intermediates. Buffers or media can be prepared in bags for midscale operations, which can be further simplified with predispensed chemicals. These applications will enable closed processing in most unit operations, and process changeover will be measured in hours instead of days as required in conventional facilities. Eliminating the testing requirements in changeover would alleviate concerns of cross-contamination for multiproduct facilities. The financial advantages would further become evident as companies would be able to own a multitude of bioreactors that could be connected together to form a larger batch size (a patent of the author) obviating the need to scale-up and validate several batch sizes and, instead, invest money in doing a good job at Process Analytical Technology (PAT) qualification.

Although disposable technologies have delivered success in development laboratories and GMP production suites, challenges and improvement opportunities yet remain. For example, concerns with extractables and leachables have not been fully resolved even though they have been addressed in detail. System integrity issues could lead to contaminations or loss of product. Product quality consistency and lot-to-lot variability present additional challenges to wider acceptance of disposable systems. The ongoing cost of disposables is always a problem to consider especially as the capital cost of hard-walled systems is fully amortized; however, the general consensus is that despite this accounting system, the disposable systems are preferred, one reason being that newer developments and improvements in disposable components will allow the manufacturer to always be using state-of-the-art technology rather than be stuck with a decades-old system.

The application of disposable technologies has changed the world of manufacturing, not only because it brings benefits to existing manufacturers but also because it may lower the cost of entry for newcomers. CMOs, manufacturers of biosimilars, and new players from emerging markets may seek the same advantages of disposable technologies and compete effectively in the global market.

Summary

Companies have to weigh a lot of issues when making decisions regarding their use of disposable technologies, some of which are economical, while others are technical or regulatory in nature. Faced with these issues, the engineers who implement new technologies at biopharmaceutical manufacturing companies often face pushback from corporate management. There are technical reasons why companies have not opted to implement disposable technologies, such as when manufacturing solutions are available for mammalian cell cultures. Disposable bags are not practical if one is using *Escherichia coli* or yeast because the biomass levels are so high that mixing, oxygenating, etc., are not easy with current disposable technology. This has, however, recently changed with sparged 2D flex reactors (www.mayabio. com), and now the choice is wide open for every type of cell or organism. There are also limitations with disposable centrifuges for harvesting the product after fermentation. Customers generally implement new technologies after they are truly tested and well proven.

Another technical issue that can delay the acceptance of disposable solutions is scale. Many of these disposable solutions cannot be scaled up to the very large scale that the manufacturers want. So, when growing mammalian cell cultures in Wave bags, the maximum scale is 500 L, which is fine, but if one wants to scale up to 2,000 L or 10,000 L, one would have to invest in much more hardware. For example, some bag solutions for bioreactors combine all the advantages, being both ready-to-use and readily disposable. They use a rocking table for mixing and require very little handling and intervention. They are ideal when one wants to make a process really lean in terms of logistics, installation, preparation, lack of cleaning, dismantling, turnaround, etc. This is certainly a dynamic area in terms of available products, but bag solutions at a larger scale (up to 2,000 L) require more hardware and handling, such as supportive bag-holders, internal mixers, and other components, making the whole system more complex and less plug and play or unplug and throw away. One loses the level of containment, and one has a lot more handling, introducing risks and being less amenable to lean approaches. However, for many applications, smaller scales are enough, and bags offer speed and flexibility beyond what steel can deliver.

Smaller companies find single-use technologies attractive because they can be set up quickly with reduced capital requirements and operated in a relatively inexpensive lab space to produce a drug under current Good Manufacturing Practices (cGMP) conditions. Single-use technologies are powerful strategic tools for smaller companies who need to advance their early stage drugs to Phase 2 or even Phase 3 clinical trials before partnering for market approval and commercial manufacturing. Single-use technologies allow promising new drug candidates to move through the approval process quickly and at reduced expense when compared to building new stainless steel facilities. One trend that is expected to drive the implementation of single-use technologies is the rise of specific drugs for smaller patient populations and personalized medicines. Combining this with the trend for higher upstream yields of an unpurified drug, a 2,000 L bioreactor could easily become the manufacturing tool of choice. In this way, single-use technologies are closing the gap between development, clinical trials, and commercial-scale manufacturing.

The Bio-Process System Alliance (BPSA) is an excellent support portal for newcomers to disposable systems and also for seasoned practitioners. The BPSA was formed in 2005 as an industry-led corporate member trade association dedicated to encouraging and accelerating the adoption of single-use manufacturing technologies in the production of biopharmaceuticals and vaccines. BPSA facilitates education, sharing of best practices, development of consensus guides, and business-to-business networking opportunities among its member companies.

Given in the following is an FAQ derived from BPSA archives:

What makes up a "typical" Single-Use System (SUS)?	SUSs consist of fluid path components to replace reusable stainless steel components. The most typical systems are made up of bag chambers, connectors, tubing, and filter capsules. For more complex unit operations such as cross-flow filtration or cell culture, the SUSs will include other functional components such as agitation systems and single-use sensors.
What are the primary benefits of SUSs?	SUSs boast improved productivity, cost structure, and a reduced environmental footprint compared to traditional stainless steel facilities. This is driven by the demanding cleanliness and sanitization standards in the biopharmaceutical industry. Productivity: Workers spend much less time changing out disposable systems, then they do cleaning and sanitizing a traditional stainless steel system. Cost: Without reusable parts to clean, there are no chemicals, water, steam, or other utilities used in the cleaning/sanitizing process. Also, a facility engineered for disposables is simplified, using less space, so the total energy consumption is reduced. Environmental footprint: While the plastics in the SUSs are usually incinerated, the footprint contributed by this is less than then that contributed by wastewater, chemicals, and energy used for cleaning traditional stainless steel systems. *Cost-effectiveness:* Single-use bioprocessing can reduce capital cost for building and retrofitting biopharmaceutical manufacturing facilities. *Manufacturing efficiencies:* SUSs can achieve significant reductions in labor, faster batch turnaround, and product changeover; SUS modularity facilitates scale-up, speeds of integration, and accelerates batch changeovers and retrofits.

What are the primary benefits of SUSs?

Quality and safety: Single-use offers reduced risks of cross-contamination, reduced risk of bioburden contamination, reduction of cleaning (and cleaning validation) issues, and other benefits that can help satisfy the requirements of regulatory agencies. *Sustainability:* Single-use bioprocesses systems can provide a range of environmental benefits beyond those of stainless steel systems. Although SUSs may generate additional solid waste, benefits include reduction in the amount of water, chemicals, and energy required for cleaning and sanitizing, as well as avoiding the labor-intensive cleaning processes required with stainless steel systems.

Are SUSs limited to specific stages of the manufacturing process?

Currently, single-use technology can be utilized in a variety of stages as well as unit operations in bioprocessing manufacturing. However, some unit operations and process scales currently have limitations due to existing capital installed and lack of technologies available in SUS formats. Historically, media and buffer preparation were some of the first-unit operations to utilize single-use technologies. As technologies have evolved, there are now larger scales and more unit operations such as cell culture, mixing, purification, formulation, and filling. As the single-use industry continues to evolve and the bioprocessing industry realizes the benefits, further investments will be required from the supply base to develop even larger scales as well as newer technologies.

Do manufacturers need to replace all stainless technology to take advantage of the benefits of SUSs?

Not necessarily—the benefits of integrating SUSs into unit operations can be achieved by making either a full conversion to plastic-based solutions or an appropriate combination of single-use and stainless technologies. Manufacturers can increase process flexibility and improve efficiencies even with partial conversion to single-use technologies. Examples include cell culture operations integrating single-use bioreactor technology with traditional stainless reactors and final fill operations combining SUSs with traditional vial-filling equipment.

Are SUSs limited to specific stages of the manufacturing process?

Currently, single-use technology can be utilized in a variety of stages as well as unit operations in bioprocessing manufacturing. However, some unit operations and process scales currently have limitations due to existing capital installed and lack of technologies available in SUS formats.

What are the factors to consider when implementing a single-use technology?

There are four basic questions/areas to ask when thinking about implementing single-use technologies: application, capital for investment, data available, and risk tolerance. What is my *application* (product and process)? Will I be performing upstream and/or downstream processing of cell therapy, drug product, or other biopharma process? What part or stage of the process will I be working? Is my process/product scope and overview well understood and documented?

What are the factors to consider when implementing a single-use technology?

Is my application mixing and buffer preparation? Is my application fermentation or other bioreactor cell growth process? Will I require significant filtration and refinement? Will I be making the final drug product? What environmental control and handling procedures do I have in place and how to they differ with single-use? How much *capital* do I have available to invest in initial start-up and/or technology conversion costs? What *data* is available from the suppliers to help answer the questions I will get from the FDA when validating my product and process? Do the suppliers have a Drug Master File (DMF)? Have the suppliers performed extensive testing on extractables/leachables, United States Pharmacopoeia (USP) Class VI, etc.? How can I reduce the amount of *risk* associated with my product/process? Am I the first to market with this process/product? Do others have a roadmap or platform that can be followed instead of reinventing the wheel? What data do I have or is available that can help reduce the amount of risk? Have I implemented QbD to eliminate future problems and issues? There are many case studies available on how companies have addressed the four topics mentioned earlier. Refer to other specific FAQ questions links or BPSA's landing page: www.bpsalliance.org. For specific inquiries and a referral to an industry expert or liaison, contact ottk@socma.com.

What effect can single-use technology have on capital and start-up costs associated with designing and commissioning a new manufacturing facility?

SUSs typically have lower capital and start-up costs. The capital equipment associated with supporting a single-use unit operation is far less expensive than for multiuse stainless systems. One example is that the materials used to build support systems for single-use operations need not be electropolished and passivated, similar to the product contact surfaces of a multiuse system. Commissioning a SUS is often *faster and simpler* than a multiuse system. The product contact surfaces or SUSs can be qualified well in advance of the actual installation, compressing start-up commissioning time lines.

How does single-use technology affect ongoing manufacturing efficiencies and production costs?

Manufacturing efficiencies and production costs will vary by company, by manufacturing site, and by unit operation. In general, overall resource deployment is less with single-use operations when compared with stainless operations. A BPSA member company expert can help you evaluate the economics and efficiencies. To be put in touch with a company expert, please contact ottk@socma.com.

Are there established standards or guides for SUSs?

While there are no industry standards specifically for single-use process systems, there are several standards and technical guides that can be applied. Some specific examples are the ASTM International Standard F838-05 on Sterilizing Filtration, the American National Standards Institute (ANSI)/American Association for the Advancement of Medical Instrumentation (AAMI)/ International Organization for Standards (ISO) Standard 11137 on Sterilization of Healthcare Products—Radiation and the ANSI/AAMI/ISO Standard 13408 on Aseptic Processing of Healthcare Products.

These and other industry standards and technical, as well as government regulations and regulatory agency guidelines, which can be applied to SUSs are further described in the BPSA Component Quality Reference Matrix. BPSA is also a leader in developing new best practice guides for SUSs that are simulating the development of standards or guides by other organizations (e.g., ASTN Bioprocessing Equipment Standard (ASTM-BPE), Parenteral Drug Association (PDA), International Society for Pharmaceutical Engineering (ISPE). BPSA guides also cover irradiation and sterilization validation, determination of extractables and leachables, and disposal of SUSs. See these at www.bpsalliance.org.

Is there information available on irradiation and sterilization/validation?

The American National Standards Institute (ANSI), American Association of Medical Instrumentation (AAMI,) and the International Organization for Standardization (ISO) have jointly issued a standard on Sterilization of Healthcare Products (ANSI/AMI/ISO 11137) that is recognized by regulative authorities around the globe. To help biopharmaceutical manufacturers and single-use equipment suppliers better understand the application of this standard and its various options to SUSs, BPSA has published a *Guide to Irradiation and Sterilization of Single-Use Systems*. This BPSA guide explains the basic principles of sterilization validation under ANSI/AAMI/ISO 11137, discusses considerations in choosing among alternate approach options, and suggests where microbial control by irradiation without validation may be applicable and beneficial in terms of development time and cost without compromising safety or quality. See the references at www.bpsalliance.org.

What are extractables and leachables, and how should they be considered during validation?

Drug developers and regulators are concerned about the potentially adverse impact on drug product quality or safety by chemicals that may migrate into the drug product from fluid contact process equipment (including SUSs), final drug product containers, or secondary packaging. Extractables are chemicals that migrate from fluid contact materials under exaggerated conditions (e.g., solvent, time, temperature) and represent a "worst-case" library of chemicals that could potentially contaminate or interact with the drug product. Leachables are chemicals that can be found in final drug product and typically include some extractables from process equipment as well as from the final container/closures or packaging along with any reaction or degradation products of those extractables and the active drug. The BPSA Guide—*Recommendations for Extractables and Leachables Testing*, provides a risk-based approach for determination of extractables and leachables from single-use process systems that has been recognized by US FDA CBER reviewers and applied successfully by several biopharmaceutical manufacturers. See the guide at www.bpsalliance.org.

What are the options for disposing of SUSs and components?

There are a range of disposal options for SUSs; the best solution will be dependent on the composite and volume of plastics, local regulations, and available waste treatment facilities. Although recycling is viewed as environmentally appealing, it is not amenable to most SUSs due to low volumes and mixed plastic content. Landfill options for typical systems include treated, untreated, as well as grind and autoclave. Incineration is a widely accepted treatment option in both the United States and Europe that reduces the volume of waste. Cogeneration is an attractive alternative that converts the plastic waste into energy that produces heat or electricity for consumption by individual facilities or entire communities. Pyrolysis is a relatively new technology that converts plastic waste into oil that can be used as fuel. To learn more about the advantages and disadvantages of each option, please refer to BPSA's *Guide to Disposal of Single-Use Bioprocess Systems* at www.bpsalliance.org.

How does the environmental impact of single-use manufacturing compare to traditional stainless manufacturing?

The total environmental footprint of a manufacturing facility is more complex and adds additional factors to the more familiar term "carbon footprint." The environmental footprint additionally considers the consumption of water, usage of land, and the impact of humans that operate and travel to, from, and within the facility. The primary contributor to the difference between traditional stainless operations and single-use operations is in sanitization and cleaning processes. While there has been growing public concern over the generation of solid waste products from plastics, by comparison the environmental burden of multiuse stainless operations is extreme in the consumption of chemicals, water, and energy.

SUSs do not require the same sanitization and cleaning rigor due to their disposability. The disposable plastic materials from single-use processes can typically be effectively incinerated often with energy capture potential in cogenerative operations.

Who can be contacted to learn more about implementing single-use technology into my operation?

BPSA comprises 40-member organizations within which reside a variety of experts who are willing to assist you in the implementation of single-use systems. The first stop to gathering the knowledge to educate on single-use manufacturing is to contact BPSA Executive Director Kevin Ott, ottk@socma.com, who can direct you to the proper resources contained within the membership of both the BPSA and the additional standard-setting and technical organizations that deal with all aspects of single-use systems.

What is the value of BPSA and the benefits of membership?

BPSA's mission is to advance the adoption of single-use systems, worldwide. BPSA promotes this purpose by conducting networking activities, monitoring legislative and regulatory initiatives involving SUS, publishing industry guides on specific topics relevant to the industry, and being the information clearinghouse for education on all aspects of single-use. Suppliers of single-use components, end-users of single-use systems, ancillary service providers and contract manufacturers are all eligible for BPSA membership as dues-paying organizations. Membership information and applications for membership can be seen at www.bpsalliance.org. Finally, BPSA is the only national organization that represents industry interests in single-use and that serves as the focal point for education on matters related to this industry and its customer base.

Appendix I: Complete Lines of Disposable Systems

Sartorius-Stedim (www.sartorius-stedim.com) offers its disposable technology factory that includes

1. FlexAct is a new system that enables one to custom-configure disposable solutions for entire biomanufacturing steps. The FlexAct system consists of the central operating module that offers the widest variety of configuration options, so one can take complete control of practically any steps in upstream and downstream processing. FlexAct CDS offers configurable disposable solutions for buffer preparation (BP), cell harvest (CH), virus inactivation (VI), media preparation (MP), and virus removal (VR). Next in line are UF DF cross-flow (UD), polishing (PO), form fill (FF), and form transfer (FT).
2. Biostat CultiBag STR Plus used for cell culture at all levels from 50 L to 1,000 L includes single-use optical dissolved oxygen (DO) and pH measurements (Figure 5.5).
3. Flexboy is a flexible bag system.
4. Flexel 3D mixing bags.
5. LevMixer and Palletank for mixing bases.

Pall Corporation (www.pall.com) also offers an extensive range of disposable products including

1. Allegro™ 2D and 3D Biocontainers
2. Allegro™ Jacketed Totes
3. Allegro™ Disposable Systems—Recommended capsule filters and membrane
4. Kleenpak™ Nova Sterilizing-grade and Virus Removal Capsule Filters
5. Kleenpak™ Sterile Connectors
6. Kleenpak™ Sterile Disconnectors
7. Kleenpak™ TFF CapsulesStax™ Disposable Depth Filter Systems
8. Stax™ Disposable Depth Filter w/ Seitz® AKS Activated Carbon Media

Millipore (www.millipore.com) offers an extensive line of complete systems. With Mobius FlexReady Solutions, one can install equipment, configure applications, and validate processes quickly and easily. Mobius FlexReady

TABLE 1.1

Pall Systems

Application/unit operation	Mixers	FlexReady solutions	CellReady bioreactor	Assemblies	Tubes and connectors	Drums	Novaseal crimper	Lynx sterile connector	Bins
Buffer and media preparation		X		X		X			
Mixing	X			X					
Bioreactor			X						
Clarification		X							
Column protection				X					
Trace contaminant Removal									
Virus removal		X							
Ultrafiltration/diafiltration		X							
Sterile filtration				X					
Storage				X		X			X
Transport				X		X			X
Sampling				X					
Connectology					X		X	X	

Solutions target key steps in the monoclonal antibody (mAb) processing/ purification train, designed and optimized for

- Clarification
- Media and buffer preparation
- Tangential flow filtration (TFF)
- Virus filtration

GE Lifesciences (www.gelifesciences.com) offer a ReadyToProcess system to include all steps of upstream and downstream processing. These components include

1. **Bioreactors**

   The Wave Bioreactor is a scalable, effective, cost-efficient rocking platform for cell culture. The culture medium and cells contact only a presterile, disposable Cellbag ensuring very short setup times. There is no need for cleaning or sterilization, providing easy operation and protection against contamination. The rocking motion of the platform induces waves in the culture fluid to provide efficient mixing and gas transfer, resulting in an environment well-suited for cell growth. The Wave system is completely scaleable across our platform ranging from 200 mL to 500 L.

2. **Disposable bioreactor bags**

   Manufactured from multilayered laminated clear plastic, Cellbag disposable bioreactors are suitable for specific cell culture process needs for research, development, or cGMP manufacturing operations. Cellbag components are similar to those used for biological storage bags and meet USP Class VI specifications for plastics. Validation data and Cellbag DMF are available to demonstrate biocompatibility. Cellbags can be highly customized to meet specific processing requirements.

3. **Fluid Management**

   ReadyCircuit assemblies comprise bags, tubing, and connectors. Together with ReadyToProcess filters and sensors, ReadyCircuit assemblies form self-contained bioprocessing modules that maintain an aseptic path and provide convenience by removing time-consuming process steps associated with conventional systems. Bags, tube sets, filters, and related equipment can be secured in appropriate orientations for efficient operation using the ReadyKart mobile processing station. With an array of features, and optional accessories, the

ReadyKart is designed to support a variety of process-specific, fluid handling needs.

4. **ReadyToProcess Konfigurator**

ReadyToProcess Konfigurator lets one design fluid handling circuits with ease online. Enter the parameters to generate the design needed, includes fast output of Piping and Instrumentation (P&ID) drawings and convenient Bill of Materials for simplified ordering.

5. **Connectivity**

ReadyMate connectors are genderless aseptic connectors that allow simple connection of components maintaining secure workflows and sterile integrity. Additional accessories, such as tube fuser and sealer of thermoplastic tubing support, secure aseptic connectivity throughout the manufacturing process.

6. **Filters**

ReadyToProcess filters are a range of preconditioned and ready-to-use cartridges and capsules for both cross-flow and normal-flow filtration operations. Factory prepared to water for injection quality for endotoxins, total organic carbon (TOC), and conductivity and sterilized via gamma radiation. They enable simpler and faster bioprocessing with maximum safety.

Chromatography Columns

ReadyToProcess columns are high-performance bioprocessing columns that come prepacked, prequalified, and presanitized. ReadyToProcess columns are designed for seamless scalability, delivering the same performance level as available in conventional processing columns such as AxiChrom and BPG.

1. **Chromatography system—ÄKTA ready**

ÄKTA ready (www.gehealthcare.com) is a liquid chromatography system built for process scale-up and production for early clinical phases. The system operates with ready-to-use, disposable flow paths and as a consequence, cleaning between products/batches and validation of cleaning procedures is not required. ÄKTA ready is a liquid chromatography system built for process scale-up and production for early clinical phases. System meets Good Laboratory Practices (GLP) and cGMP requirements for Phases I–III in drug development and full-scale production, provides improved economy and productivity due to simpler procedures, single-use eliminates risk of cross-contamination between products/batches, easy connection to and operation with prepacked ReadyToProcess columns, and other process columns, scalable processes using UNICORN software.

2

Safety of Disposable Systems

Out of this nettle, danger, we pluck this flower, safety.

William Shakespeare

Disposable devices from filter housings to the lining of bioreactors make extensive use of plastic materials or elastomer systems. Today, perhaps the most significant impediment in the wider acceptance of disposable systems is the controversy surrounding the possibility of contaminating the product from the chemicals in the plastic film. So, before entering a broad description of the choices of disposables available, this topic should be examined in detail.

> All final containers and closures shall be made of material that will not hasten the deterioration of the product or otherwise render it less suitable for the intended use. All final containers and closures shall be clean and free of surface solids, leachable contaminants and other materials that will hasten the deterioration of the product or otherwise render it less suitable for the intended use. (Biologics *21CFR600.11(h)*)

Leachables are chemicals that migrate from disposable processing equipment into various components of the drug product during manufacturing. Extractables are chemical entities (organic and inorganic) that can be extracted from disposables using common laboratory solvents in controlled experiments. They represent the worst-case scenario and are used as a tool to predict the types of leachables that may be encountered during pharmaceutical production.

The issue of chemicals leaching from plastic has been the hottest topic not just for the bioprocess industry but also for many other industries including the food industry, where issues such as the safety of bisphenol-A (BPA) in water bottles keep rising. A Google search of the topic results in millions of hits. How the use of plastic affects bioprocessing is of great interest to the stainless steel industry.

While regulatory requirements pertain to the toxic effects of leachables, a risk unique to biological drugs arises in the effect of leachables on the three- and four-dimensional structure of protein drugs: such changes can render the drug more immunogenic if not less effective, and these side effects are thus of greater importance to the bioprocessing industry. The most well-known problem is the high incidence of pure red cell aplasia (PRCA) reported in patients using commercial erythropoietin formulations leading to several deaths. While the source of the PRCA induction is not clearly settled, it is generally attributed to a change in the drug formulation that included a new surfactant, which caused unexpected leaching of an elastomer compound from the rubber stopper.

Polymers and Additives

The materials used to fabricate single-use processing equipment for biopharmaceutical manufacturing are usually polymers, such as plastic or elastomers (rubber), rather than the traditional metal or glass. Polymers offer more versatility because they are lightweight, flexible, and much more durable than their traditional counterparts. Plastic and rubber are also disposable, so issues associated with cleaning and validation can be avoided. Additives can also be incorporated into polymers to give them clarity of glass or to add color to labels or to code parts.

Unlike metals, where the risk lies mainly in oxidation, polymers are affected by heat, light, oxygen, and autoclaving and, thus, degrade over time if not stabilized, and this can adversely affect the mechanical properties. Polymers are thus stabilized by incorporating chemicals that are prone to leaching during the manufacturing process and storage of biological materials.

When a plastic resin is processed, it is often introduced into an extruder, where it is melted at high temperatures and its stability is influenced by its molecular structure, the polymerization process, the presence of residual catalysts, and the finishing steps used in production. Processing conditions during extrusion (e.g., temperature, shear, and residence time in the extruder) can dramatically affect polymer degradation. End-use conditions that expose a polymer to excessive heat or light (such as outdoor applications or sterilization techniques used in medical practices) can foster premature failure of polymer products as well, leading to a loss of flexibility or strength. If left unchecked, the results often can be the total failure of the plastic component.

Polymer degradation is controlled by the use of additives, which are specialty chemicals that provide a desired effect on the polymer. The effect can be stabilization, which allows a polymer to maintain its strength and flexibility or performance improvement, which adds color or some special characteristic such as antistatic or antimicrobial properties. There are typically three classes of stabilizers:

- Melt processing aids such as phosphites and hindered phenols, antioxidants that protect a polymer during extrusion and molding

- Long-term thermal stabilizers that provide defense against heat encountered in end-use applications (e.g., hindered phenols and hindered amines)
- Light stabilizers that provide ultraviolet (UV) protection through mechanisms such as radical trapping, UV absorption, or excited-state quenching

One application in which an additive can improve or alter the performance of a polymer is a filler or modifier that affects its mechanical properties. Additives known as *plasticizers* can affect the stress–strain relationship of a polymer. Polyvinylchloride (PVC) is used for home water pipes and is a very rigid material. With the addition of plasticizers, however, it becomes very flexible and can be used to make intravenous (IV) bags and inflatable devices. Lubricants and processing aids are also used to reduce polymer manufacturing cycle times (e.g., mold-release agents) or facilitate the movement of plastic and elastomeric components that contact each other (e.g., rubber stoppers used in syringes).

Additives are not always single entities. Some are manufactured from naturally occurring raw materials such as tallow and vegetable oils that are themselves composed of many different components and can vary from batch to batch. Others are considered "products by process," as they are formed during processing by adding several starting materials to affect the chemical reaction. The complexity of chemical reactions that take place in the manufacturing of plastic makes the analysis of extractables and leachables very complex and difficult. In testing extractables and leachables, those lesser-known minor chemical species may be the ones that leach into a drug product, but this is not predictable as it is to a greater degree a function of the characteristics of the product.

Stabilizers incorporated into plastics and rubbers are constantly working to provide much-needed protection to the polymer substrate. This is a dynamic process that changes according to the external stress on the system. For example, good stabilizers are efficient radical scavengers. Generally, a two-tiered approach is used to protect polymers from the heat and shear they encounter during processing: using primary antioxidants (e.g., hindered phenols such as butylated hydroxy toluene (BHT) or Ciba's Irganox) for protection during processing can provide long-term heat stability. Secondary antioxidants are also added as process stabilizers, typically hydroperoxide decomposers that protect polymers during extrusion and molding and protecting the primary antioxidants against decomposition. All the by-products of these reactions become available to leach from polymers into a drug product.

Elastomers are also used for special stabilization: acid scavengers are used to neutralize traces of halogen anions formed during the aging of halogen-containing rubbers (e.g., brominated or chlorinated isobutylene isoprene). If not neutralized, anions cause premature aging and a decrease in the performance of rubber articles over time. Metal oxides can be very efficient acid scavengers. Ions of copper (Cu), iron (Fe), cobalt (Co), nickel (Ni), and other transition metals that have different oxidation states with comparable stability are called *rubber poisons* because they are easily oxidized or reduced by one-electron transfer. They are very active catalysts for hydroperoxide decomposition and contribute to the degradation of rubber vulcanizates. Rubber poisons thus requires a specific stabilizer: a metal deactivator, such as 2,3-bis[[3-[3,5-di-*tert*-butyl-4-hydroxyphenyl]propionyl]]propionohydrazide, which binds ions into stable complexes and deactivates them.

As a result of the need to add chemicals to elastomer systems, the extractables in a disposable system can include

- Monomer and oligomers from incomplete polymerization reactions
- Additives and their transformation and degradation products
- Lubricants and surface modifiers
- Fillers
- Rubber curing agents and vulcanizates
- Impurities and undesirable reaction products such as polyaromatic hydrocarbons, nitrosamines, and mercaptobenzothiazoles

Unexpected additives can also be present in a polymer system because of the inconsistencies in the process of manufacturing elastomer systems whereby unpredictable reactions can take place.

Despite the risk in the use of additives added to polymers, the utility of polymers in disposable bioprocess equipment (and in all medical or pharmaceutical applications) far outweighs the risks associated with their use. These risks can be managed well by taking three steps: material selection, implementation of a proper testing program, and partnering with vendors.

Material Selection

The type of plastic used should match the needed physical and chemical properties and compatibility of the additives used for the product manufactured. For example, phenolic antioxidants, each with the same active site (the hindered phenol moiety) but with different nature of the remainder of the molecule, make them soluble or compatible with a given polymer substrate. An antioxidant that is compatible with nylon might not be the best choice for use in polyolefins, as an example.

Ensuring compatibility often lessens the amount of leaching that can occur. It is also important to select polymers and additives that are approved for use by the regulatory authorities for the specific use. Such compounds have already undergone a fair amount of analytical and toxicological testing, so a good amount of information is often available for them. Because of this, most manufacturers are likely to continue using these additives, and thus, the user may not have to alter the composition at a later stage as these compounds and the art of using them is likely to survive obviating the need for a change control step as significant changes in the process need to be reported back to the FDA.

Commercially supplied plastic films are proprietary formulations and arrangements; for example, Advanced Scientific produces its bags utilizing two films. The fluid contact film is a 5.0 mm polyethylene. The outer is a 5-layer 7 mil co-extrusion film that provides a barrier and durability. A typical test report is given in Table 2.1.

ATMI offers its proprietary TK8 film, which is constructed from laminated layers of polyamide (PA), ethylene vinyl alcohol polymer (EVOH), and ultralow density polyethylene (ULDPE). The outer PA layer provides robust puncture resistance, strength, and excellent thermal stability. The EVOH layer minimizes gas diffusion across the film while maintaining a very good flex crack resistance. The ULDPE layers provide flexibility, integrity, and an ultra-clean, ultrapure, low-extractables product-contacting layer. The combination of these layers results in a film that has outstanding optical clarity is easy to handle and performs well in a broad range of bioprocess applications. The inner ULDPE layer used in TK8 is blow-extruded in-house by ATMI under cleanroom conditions (0.2 μm filtered air), ensuring the cleanest possible product-contacting surface. Lamination is also performed under controlled, ultra-clean conditions. Lastly, TK8 film is converted into ATMI bag products in an ISO Class 5 cleanroom. All of the layers in TK8 are made from medical-grade materials, meaning that they comply with industry standards and are subject to strict change controls. The entire structure of TK8 is totally free of any animal-derived components (ADCF). ATMI has also created TK8 with dual sourcing and contingency planning in mind, to ensure security of supply.

- TK8 film complies with USP Class VI (USP 87, USP 88, and USP 661).
- ULDPE resin complies with EP 3.1.3.
- Shelf life is supported by aging validation studies.
- Certified ADCF.
- Bioburden evaluation available (ISO 11737).
- Particle count data available (EP 2.9.19 or USP 788).
- By performing blow extrusion in-house, ATMI maintains full control and traceability of the contact film composition, from resin to finished bag product.

Testing

Polymers used in medical and pharmaceutical applications should comply with the appropriate USP guidelines, and it is recommended that they meet

TABLE 2.1

Initial Evaluation Tests for Consideration

Device Categories			Biological Effect							
Body Contact (see 4.1)		Contact Duration (see 4.2) A—Limited (24 h) B—Prolonged (24 h to 30 days) C—Permanent (>30 days)	Cytotoxicity	Sensation	Irritation or Intracutaneous Reactivity	System Toxicity (acute)	Subchronic Toxicity (subacute toxicity)	Genotoxicity	Implantation	Hemocompatibility
Surface Devices	Skin	A	x	x	x					
		B	x	x	x					
		C	x	x	x					
	Mucosal Membrane	A	x	x	x					
		B	x	x	x	o	o		o	
		C	x	x	x	o	x	x	o	
	Breached or Compromised Surfaces	A	x	x	x	o				
		B	x	x	x	o	o		o	
		C	x	x	x	o	x	x	o	
External Communicating Devices	Blood Path, Indirect	A	x	x	x	x				x
		B	x	x	x	x	o			x
		C	x	x	o	x	x	x	o	x
	Tissue/Bone/ Dentin Communicating+	A	x	x	x	o				
		B	x	x	o	o	o	x	x	
		C	x	x	o	o	o	x	x	
	Circulating Blood	A	x	x	x	x		o^		x
		B	x	x	x	x	o	x	o	x
		C	x	x	x	x	x	x	o	x
Implant Devices	Tissue/Bone	A	x	x	x	o				
		B	x	x	o	o	o	x	x	
		C	x	x	o	o	o	x	x	
	Blood	A	x	x	x	x			x	x
		B	x	x	x	x	o	x	x	x
		C	x	x	x	x	x	x	X	x

Note: X = ISO Evaluation Tests for consideration.
O: Additional tests that may be applicable.
+: Tissue includes tissue fluids and subcutaneous spaces.
^: For all devices used in extracorporial circuits.

USP Class VI testing as documented in USP 88. Appropriate extractables and leachables testing programs must be implemented for all bioprocessing materials that come into direct contact with the drug.

The best practice guidelines for conducting such testing are provided by the Bio-Process System Alliance (BPSA) as a two-part technical guideline for evaluating the risk associated with extractables and leachables, specifically for single-use processing equipment. This organization is dedicated to encouraging the use of disposable systems and provides excellent support and assistance; the reader is highly encouraged to visit their website for newer information, as well as participate in their many seminars and conventions to stay abreast of the developments in this fast-changing field.

The testing for leachables should not necessarily end once the materials have been qualified. It is necessary to have in place a quality control program instead of testing the product or equipment alone. The level of quality control testing will depend on risk tolerance. Fortunately, the manufacturing of recombinant drugs involves extensive purification steps that are likely to remove most of these leachables. Also, the final medium used for protein solutions is aqueous, and many of the leachables are not soluble in water: this further reduces the risk. Greater risk can be seen in the final packaging components; for example, rubber stoppers used in packaging the final dosage form are more likely to be a risk to the protein formulation than any other component in the chain of disposables to which the drug is exposed during the manufacturing process.

While it is always a good idea to establish in-house testing of leachables, often it is neither possible nor recommended; several highly reputed laboratories have fully certified programs; the following is a short list of these laboratories:

Product Quality Research Institute (www.pqri.org)

Rapra Technology, Ltd. (www.rapra.net)

Impact Analytical (www.impactanalytical.com)

Avomeen Analytical Services (www.avomeen.com)

SGS North America (www.us.sgs.com)

Cyanta (www.cyanta.com)

Irvine Pharmaceutical Services (www.ialab.com)

Pace Analytical Services, Inc. (www.pacelabs.com)

American Society for Quality (www.asq.org)

Intertek (www.intertek.com)

Partnering with Vendors

Reputable vendors often have extractables data already on hand to share with customers. In many cases, they will provide a certificate of analysis and toxicological information associated with materials used to fabricate their products. Vendors also should have well-established change control processes for the products they sell to allow sponsors to modify their applications with regulatory agencies accordingly.

Responsibility of Sponsors

Companies filing regulatory approval have the responsibility of complying with the requirements of validating the process to minimize the risk from leachables. Extractables and leachables evaluations are part of a validation program for processes using disposable biopharmaceutical systems and components. There is minimal regulatory guidance that directly addresses extractables and leachables in bioprocessing.

The extractables are evaluated by exposing components or systems to conditions that are more severe than normally found in a biopharmaceutical process, typically using a variety of solvents at high temperatures. The goal of an extractable study is to identify as many compounds as possible that have the potential to become leachables. A positive outcome is one where the list of extractables from a material is sizable. Although it is not expected that many of those extractables will actually leach into the drug product at detectable levels, a materials extractables profile provides critical information in pursuit of a comprehensive leachables test.

Not all leachables may be found during the extractables evaluation because drug formulation components or buffers may interact with a polymer or its additives to form a new "leachable" contaminant that was not previously identified during the extractables analysis. In addition, leachables that were not identified as extractables also will be found if the drug product formulation and processing conditions are unique and more severe than the conditions at which extractable tests were performed—or when the analytical methodologies used in the two types of studies are different.

Regulatory Requirements

There are as yet no specific standards or guidance that reference extractables and leachables from disposable bioprocessing materials. Many references that do apply were written to address the processing materials and equipment without regard to the materials of construction.

United States and Canada

The foundation for the requirement to assess extractables and leachables in the United States is introduced in Title 21 of the Code of Federal Regulations (CFR) Part 211.65, which states that

> Equipment shall be constructed so that surfaces that contact components, in-process materials, or drug products shall not be reactive, additive, or absorptive so as to alter the safety, identity, strength, quality, or purity of the drug product beyond the official or other established requirements.

This regulation applies to all materials, including metals, glass, and plastics.

Extractables and leachables generally would be considered "additive," although it is also possible for leachables to interact with a product to yield new contaminants.

The U.S. FDA regulatory guidance for final container–closure systems, though not written for process-contact materials, gives directions about the type of final product testing that may be provided regarding extractables and leachables from single-use process components and systems. The May 1999 guidance document from the FDA's Center for Drug Evaluation and Research (CDER) indicates the types of drug products and component dosage form interactions that the FDA considers to be the highest risks for extractables. Generally, the likelihood of the packaging component interacting with dosage form is the highest in injectable dosage forms, mainly because of the low level of leachables that can be allowed in such drug delivery systems.

Drugs that will be administered as injectables or inhalants will have higher levels of regulatory concern than oral or topical drugs. Similarly, liquid dosage forms will have higher regulatory concern than tablets because extractables migrate into liquids more easily than into solids.

In addition, pharmaceutical-grade materials are expected to meet or exceed industry and regulatory standards and requirements such as those listed in USP 87 and 88. The USP procedures test the biological reactivity of mammalian cell cultures following contact with polymeric materials. Those chapters are helpful for testing the suitability of plastics for use in fabricating a system to process parenteral drug formulations. However, they are not considered sufficient regulatory documentation for extractables and leachables because many toxicological indicators are not evaluated, including subacute and chronic toxicity along with evaluation of carcinogenic, reproductive, developmental, neurological, and immunological effects.

Europe

In the European Union, a related statement to the US 21 CFR 211.65 is found in the rules governing manufacture of medicinal products. The EU good manufacturing practice document states: "Production equipment should not present any hazard to the products. The parts of the production equipment that come into contact with the product must not be reactive, additive or absorptive to such an extent that it will affect the quality of the product and thus present any hazard."

The European Medicines Evaluation Agency (EMEA) published a guideline on plastic immediate packaging materials in December 2005 that also addresses container–closure systems and has been used to provide direction for single-use process-contact materials. Data to be included relating to extractables and leachables come from extraction studies ("worst-case leachables"), interaction studies, migration studies (similar to leachable information for those components), identify what additional information or testing is required, and then set and execute a plan to fill in the gaps.

Risk Assessment

Risk assessment is based on the following considerations:

- *Compatibility of materials:* Most biological drugs formulations are aqueous based, and therefore compatible with the materials used in most disposable processing components. Still, a check to make sure that the process stream and/or formulation do not violate any of the manufacturer's recommendations for chemical compatibility, pH, and operating pressure/temperature is warranted before proceeding. A full analysis of data generated by the vendor should be completed upfront as a preparatory step.

- *Proximity of a component to the final product:* Product contact immediately before the final fill increases the risk of leachables in a final product. For example, tubing or connectors used to transfer starting buffers probably present a lower risk because of their upstream location. Processing steps such as diafiltration or lyophilization that could remove leachables from a process should also be considered because they may reduce associated risk. However, it cannot be assumed that a step that can potentially remove some leachables will remove all leachables. In such cases, supporting data should be obtained.

- *Product composition:* In general, a product stream or formulation that has higher levels of organics, particularly high or low pH, or solubilizing agents such as surfactants (detergents), will increase the regulatory and safety concern for potential leachables. Neutral buffers lower concern about potential leachables.

- *Surface area:* The surface area exposed to a product stream varies widely. It is relatively high for filters, in which the internal surface area is 1,000× the filtration area. Conversely, surface area is relatively small for O-ring seals.
- *Time and temperature:* Longer contact times allow for more potential leachables to be removed from a material until equilibrium is reached. Higher temperatures lead to more rapid migration of leachables from materials into a process stream or formulation.
- *Pretreatment steps:* Sterilization by steam autoclave and/or gamma irradiation may cause higher levels of extractables and leachables depending on the polymer formulation involved in a single-use component. On the other hand, rinsing may lower the concern for extractables and leachables (e.g., when filters are flushed before use).

Here are some highlights relating to risk assessment of extractables and leachables:

1. Regulatory responsibility for overall assessment and understanding of a finished product and process components involved in its production remains with the product sponsor. This includes evaluations of extractables and leachables. Regulatory agencies do not have a guideline available yet to help sponsors.
2. All elastomeric and plastic-based materials contain extractables specific to the formulated and cured materials from which they are constructed.
3. Contaminants are also found in stainless steel systems in the form of residues left after cleaning or traces of metals such as iron, nickel, and chromium salts from the stainless steel itself, so the problem of contamination from the container is not restricted to disposable containers.
4. Most polymers without certain additives would not work as materials of use in disposable processing: this includes stabilizing the polymer, extruding it, and preventing its oxidation and UV degradation; other additives include antistatic agents, impact modifiers, catalysts, release agents, colorants, brighteners, bactericides, and blowing agents. The choice of polymer or method of polymerization (by heat or chemical means) directly affects the levels and types of compounds found as extractables.
5. Fluoropolymers offer the best choice as they are typically processed without additives, stabilizers, or processing aids.
6. A Drug Master File (DMF) or Biological Drug Master File (BMF) for process-contact equipment is not explicitly required by U.S. regulatory authorities. However, it represents a way for vendors to share

proprietary information about a component or raw material with the FDA and to ensure that such information remains up to date. It is therefore important that sponsors work only with the most reputable suppliers of disposable components.

7. The levels and types of compounds found as extractable analytes are directly affected by the type and degree of sterilization performed (e.g., gamma irradiation, ethylene oxide gas, or autoclaving). The leachable analyte and concentration that may be of issue to one particular drug formulation may have no impact on another. It is for this reason that it is the responsibility of product sponsors to qualify and demonstrate the applicability of process components within their manufacturing systems. Leachables are always final-product specific.

8. All component materials should be evaluated that have the potential to come into direct contact with a manufactured drug product. Of greater importance are the components that would contact the product in the postpurification stage.

9. Controlled extraction studies are designed to generate extractables: the presence of extractables is expected. This does not necessarily reflect the degree and concentration of leachables that will be found upon contact with a product stream: leachables are a result of the nature of the product, the length of exposure, and the environmental conditions for the storage of the product.

10. Detection of a toxic or otherwise undesirable extractable under aggressive conditions requires testing to ensure that migration to the product is below acceptable limits under actual processing conditions. It is done by controlled extraction studies using multiple solvents of varying polarity to fully elucidate the extractable analytes in question. Techniques such as Soxhlet extraction, solvent refluxing, microwave extraction, sonication, and/or acid washing at an elevated temperature may also be used. For extractables testing, the contact surface area can be maximized by mechanical methods such as cutting or grinding.

11. For leachables testing, it is most applicable to mimic actual process conditions by leaving test components intact. Controlled extraction studies should use extraction media of varying polarities and physical properties. Ideally, this would come from using two or three solvents that include analysis by HPLC, GC-MS, and ICP-MS.

12. Toxicology of leachables should be performed using approved protocols. A Product Quality Research Institute (PQRI) document on extractables and leachables suggests an approach to address toxicology using LD50 with a 1,000× or 10,000× safety factor based on the dosage quantity. In addition, several structure–activity relationship (SAR) databases are readily available to professional toxicologists.

Examples include the "Carcinogenic Potency Database" (CPDB, http://potency.berkeley.edu/) and the U.S. Environmental Protection Agency's "Distributed Structure-Searchable Toxicity Network" (DSSTOX, http://www.epa.gov/ncct/dsstox/) database. Chances are that most sponsors will not be able to conduct these studies in-house, and it is advised to outsource these evaluations.

13. The classes of compounds that extractables include, more particularly, *n*-nitrosamines, polynuclear aromatics (sometime termed *polyaromatic hydrocarbons*, PAH), and 2-mercaptobenzothiozole, along with biologically active compounds such as bisphenol-A (BPA). Individual extractable compounds are too numerous to list, but examples include aromatic antioxidants such as butylated hydroxytoluene (BHT), oleamide, bromide, fluoride, chloride, oleic acid, erucamide, eicosane, and stearic acid. Databases on extractables are widely available, such as by PQRI. Comprehensive extractable data for components can reduce the time and resources needed to qualify leachables from the systems where they are used. When comparing supplied extractables data for components constructed of similar materials, end-users should carefully review the methods used to generate the data. Less rigorous methods may underrepresent the actual levels and extent of extractables, and a report describing more extractables may simply come from using more rigorous methods.

14. For determination of leachables in products, it is currently the industry standard to validate analytical methods according to ICH and USP criteria. This ensures appropriate levels of analytical precision and accuracy.

15. The overall quantity of extractables or leachables can be estimated using nonspecific methods such as total organic carbon (TOC) and nonvolatile residue (NVR) analysis. Such nonspecific quantitation is especially useful in comparing materials before their final selection for a process. These analyses can be used individually or collectively to estimate amounts of extractable material present and to ensure that targeted methods are not missing a major extractable constituent. For instance, nonpolar compounds without chromophores can be identified using Fourier-transform infrared (FTIR) analysis of nonvolatile residues.

16. Organic extractables will leach into formulations at a higher level if the products have higher organic content or if surfactants are present.

17. The toxicity of leachables is frequently estimated based on the amount entering the human body in each dose. Thus, it is often not the quantity of leachables in a product but how much finds its way into the human body. This is somewhat analogous to the limits many regulatory agencies set on residual DNA in a finished product.

18. The component used for an extractables study should be the same one that will be used in a process, and it should have the same pretreatment steps as is intended for that process. For instance, if a process uses gamma irradiation for sterilization, then the component used for extractables testing should be sterilized by this method. Often, a vendor will provide simulated data based on similar products by extrapolating the data from other components; this would not be acceptable.

19. The solvents used for leachables studies should include water and a low-molecular-weight alcohol such as ethanol or *n*-propanol. Where appropriate, an organic solvent with the appropriate solubility parameters will help identify additional extractables. Extractions should be performed at relatively extreme time and temperature conditions. However, the solvents or extraction conditions should not be so extreme as to degrade materials to a point at which they are not mechanically functional (e.g., melting or dissolving). Extreme conditions used should be relative to those under which a material is normally used. For example, one normally used at room temperature might be extracted at an elevated temperature of 50°C or 70°C.

20. Analytical methods should include HPLC and GC-MS methods to detect and identify specific, individual, extractable compounds. HPLC with an ultraviolet (HPLC-UV) or mass spectrometer (LC-MS) detector and GC-MS are the most scientifically robust methods for this purpose. When metals are a concern, inductively coupled plasma analysis is widely used, both with and without mass-spectrometric detection (ICP and ICP-MS).

21. While it is desirable to identify each extractable, for some extractables, such as siloxanes and oligomers of base polymers, precise identification is not feasible because of the large number of closely related isomers and oligomers. In such cases, a general classification can be used. Quantitation of identified extractables is informative, but it does not need to be performed at a high level of precision. This is different from recommendations for evaluating extractables for final containers or closures, for which analytical and toxicological limits should be set based on a measured level of extractables.

22. User-specific components, such as filters, connecters, tubing and bags, etc., may be built by using subcomponents from different vendors. It is unlikely that the composite system would have complete data on extractables from the vendor assembling the component. Individual data for each subcomponent can be pooled, but it may be easier for the sponsor to conduct the study on the entire component at one time. It is therefore advisable that sponsors use off-the-shelf products where possible.

23. Biocompatibility testing is a very complex issue. It is a material's lack of interaction with living tissue or a living system by not being toxic, injurious, or physiologically reactive, and not causing an immunological rejection. This testing is required, and two common test regimens are commonly used to measure biocompatibility: USP 88, Biological Reactivity Testing (USP Class VI), and ISO 10993, Biological Evaluation of Medical Devices, which has replaced the USP Class VI test.

24. The ISO 10993 has 20 parts and provides testing requirements in great detail. These parts include

   a. Evaluation and testing (see Appendix I)
   b. Animal welfare requirements
   c. Tests for genotoxicity, carcinogenicity, and reproductive toxicity
   d. Selection of tests for interactions with blood
   e. Tests for in vitro cytotoxicity
   f. Tests for local effects after implantation
   g. Ethylene oxide sterilization residuals
   h. Clinical investigation of medical devices
   i. Framework for identification and quantification of potential degradation products
   j. Tests for irritation and delayed-type hypersensitivity
   k. Tests for systemic toxicity
   l. Sample preparation and reference materials
   m. Identification and quantification of degradation products from polymeric medical devices
   n. Identification and quantification of degradation products from ceramics
   o. Identification and quantification of degradation products from metals and alloys
   p. Toxicokinetic study design for degradation products and leachables
   q. Establishment of allowable limits for leachable substances
   r. Chemical characterization of materials
   s. Physicochemical, morphological, and topographical characterization of materials
   t. Principles and methods for immunotoxicology testing of medical devices

25. The USP 88 protocols are used to classify plastics in Classes I–VI, based on end use, type, and time of exposure of human tissue to

plastics, of which Class VI requires the most stringent testing of all the six classes. These tests measure and determine the biological response of animals to the plastic by either direct or indirect contact, or by injection of the specific extracts prepared from the material under test. The tests are described as

a. Systemic toxicity test to determine the irritant effect of toxic leachables present in extracts of test materials

b. Intracutaneous test to assess the localized reaction of tissue to leachable substances

c. Implantation test to evaluate the reaction of living tissue to the plastic

The extracts for the test are prepared at one of three standard temperatures/times: 50°C (122°F) for 72 h, 70°C (158°F) for 24 h, 121°C (250°F) for 1 h.

26. Typical testing data for disposable bioreactors (as supplied by GE Healthcare) would include

- Testing is performed on irradiated film (50 kGy):
- USP XXII plastic class VI and ISO 10993:
- ISO 10993-4 Hemolysis study in vivo extraction method
- ISO 10993-5 Cytotoxicity study using ISO elution method
- ISO 10993-6 Muscle implantation study in rabbit
- ISO 10993-10 Acute intracutaneous reactivity study in rabbit
- ISO 10993-11 Acute systemic toxicity in mouse

Appendix I: Use of International Standard ISO-10993 "Biological Evaluation of Medical Devices Part 1: Evaluation and Testing"

Background

The biological evaluation of medical devices is performed to determine the potential toxicity resulting from contact of the component materials of the device with the body. The device materials should not either directly or through the release of their material constituents: (i) produce adverse local or systemic effects; (ii) be carcinogenic; or (iii) produce adverse reproductive and developmental effects. Therefore, the evaluation of any new device intended for human use requires data from systematic testing to ensure that the benefits provided by the final product will exceed any potential risks produced by device materials. When selecting the appropriate tests for the biological evaluation of a medical device, one must consider the chemical characteristics of device materials and the nature, degree, frequency, and duration of their exposure to the body. In general, the tests include acute, subchronic, and chronic toxicity; irritation to skin, eyes, and mucosal surfaces; sensitization; hemocompatibility; genotoxicity; carcinogenicity; and effects on reproduction including developmental effects. However, depending on varying characteristics and intended uses of devices as well as the nature of contact, these general tests may not be sufficient to demonstrate the safety of some specialized devices. Additional tests for specific target organ toxicity, such as neurotoxicity and immunotoxicity, may be necessary for some devices. For example, a neurological device with direct contact with brain parenchyma and cerebrospinal fluid (CSF) may require an animal implant test to evaluate its effects on the brain parenchyma, susceptibility to seizure, and effects on the functional mechanism of choroid plexus and arachnoid villi to secrete and absorb CSF. The specific clinical application and the materials used in the manufacture of the new device determine which tests are appropriate. Some devices are made of materials that have been well characterized chemically and physically in published literature and have a long history of safe use. For the purposes of demonstrating the substantial equivalence of such devices to other marketed products, it may not be necessary to conduct all the tests suggested in the FDA matrix of this guidance. FDA reviewers are advised to use their scientific judgment in determining which tests are required for the demonstration of substantial equivalence under section 510(k). In such situations, the manufacturer must document the use of a particular material in a legally marketed predicate device or a legally marketed device with comparable patient exposure.

International Guidance and Standards

In 1986, the FDA, Health and Welfare Canada, and Health and Social Services UK issued the Tripartite Biocompatibility Guidance for Medical Devices. This Guidance has been used by FDA reviewers, as well as by manufacturers of medical devices, in selecting appropriate tests to evaluate the adverse biological responses to medical devices. Since that time, the International Standards Organization (ISO), in an effort to harmonize biocompatibility testing, developed a standard for biological evaluation of medical devices (ISO 10993). The scope of this 12-part standard is to evaluate the effects of medical device materials on the body. The first part of this standard, "Biological Evaluation of Medical Devices: Part 1: Evaluation and Testing," provides guidance for selecting tests to evaluate the biological response to medical devices. Most of the other parts of the ISO standard deal with appropriate methods to conduct the biological tests suggested in Part 1 of the standard. ISO 10993, Part 1, and the FDA-modified matrix thereof, use an approach to test selection that is very similar to the currently used Tripartite Guidance, including the same seven principles. It also uses a tabular format (matrix) for laying out the test requirements based on the various factors discussed earlier. The matrix consists of two tables. See Table 2.1—Initial Evaluation Tests for Consideration and Table 2.2—Supplementary Evaluation Tests for Consideration. Table 2.3 is a biocompatibility flow chart for the selection of toxicity tests for 510(k) s. It may be applicable to some Pharmaceutical Manufacturers Associations (PMAs) also but not all PMAs. In addition, the FDA is in the process of preparing toxicology profiles for specific devices. These profiles will assist in determining appropriate toxicology tests for these devices. To harmonize biological response testing with the requirements of other countries, the FDA will apply the ISO standard, Part 1, in the review process in lieu of the Tripartite Biocompatibility Guidance. The FDA notes that the ISO standard acknowledges certain kinds of discrepancies. It states: "due to diversity of medical devices, it is recognized that not all tests identified in a category will be necessary and practical for any given device. It is indispensable for testing that each device shall be considered on its own mertis: additional tests not indicated in the table may be necessary." In keeping with this inherent flexibility of the ISO standard, the FDA has made several modifications to the testing required by ISO 10993, Part 1. These modifications are required for the category of surface devices permanently contacting mucosal membranes (e.g., Intra Uterine Device (IUDs)). The ISO standard would not require acute, subchronic, and chronic toxicity and implantation tests. Also, for externally communicating devices with prolonged and permanent contact with tissue, bone, or dentin (e.g., filling materials and dental cements), the ISO standard does not require irritation, systemic toxicity, and acute, subchronic, and chronic toxicity tests. Therefore, the FDA has included these types of tests in the matrix. Although several tests were added to the matrix, reviewers should note that some tests are commonly requested, while other tests are to be considered and only asked for on a case-by-case basis. Thus, the modified matrix is only a framework for the selection of tests and not a checklist of every required test.

Reviewers should avoid a proscriptive interpretation of the matrix. If a reviewer is uncertain about the applicability of a specific type of test for a specific device, the reviewer should consult toxicologists in the Office of Device Evaluation (ODE). The FDA expects that manufacturers will consider performing the additional tests for certain categories of devices suggested in the FDA-modified matrix. This does not mean that all the tests suggested in the modified matrix are essential and relevant for all devices. In addition, device manufacturers are advised to consider tests to detect chemical components of device materials which may be pyrogenic. ISO 10993, Part 1, and the appropriate consideration of additional tests suggested by knowledgeable individuals will generate adequate biological data to meet the FDA's requirements. Reviewers in the ODE will accept data developed according to ISO-10993, Part 1, with the FDA-modified matrix as modified. Manufacturers are advised to initiate discussions with the appropriate review division in

TABLE 2.2

Supplementary Evaluation Tests for Consideration

Device Categories		Contact Duration (see 4.2) A—Limited (24 h) B—Prolonged (24 h to 30 days) C—Permanent (>30 days)	Biological Effect			
Body Contact (see 4.1)			Chronic Toxicity	Carcinogenicity	Reproductive Developmental	Biodegradable
Surface Devices	Skin	A				
		B				
		C				
	Mucosal Membrane	A				
		B				
		C	o			
	Breached or Compromised Surfaces	A				
		B				
		C	o			
External Communicating Devices	Blood Path, Indirect	A				
		B				
		C	x	x		
	Tissue/Bone/ Dentin Communicating+	A				
		B				
		C	o	x		
	Circulating Blood	A				
		B				
		C	x	x		
Implant Devices	Tissue/Bone	A				
		B				
		C	x	x		
	Blood	A				
		B				
		C	x	x		

Note: X = ISO Evaluation Tests for consideration.
O: Additional tests that may be applicable.

TABLE 2.3

Biocompatibility Flowchart for the Selection of Toxicity Tests for 510(k)s

USP Acute Systemic Injection Test	Pass	USP <88>
USP Intracutaneous Injection Test	Pass	USP <88>
USP Intramuscular Implantation Test	Pass	USP <88>
USP MEM Elution Method	Non-Cytotoxic	USP <87>
Physiochemical Test for Plastics	Pass	USP <661>

Extractables

	TOC After 90 Days (ppm)	pH Shift After 90 Days
Purified Water (pH=7)	<2	-0.79
Acidic Water (pH<2)	<3	+0.01
Basic Water (pH>10)	<4	+0.87

Physical Data

	Average Force	Average MOE	Average Elongation	
Water Vapor Transmission Rate (g/100in2/24hrs)	0.017	ASTM F-1249		
Carbon Dioxide Transmission Rate (cc/100in2/24hrs)	0.129	ASTM F-2476		
Oxygen Transmission Rate (cc/100in2/24hrs)	0.023	ASTM F-1927		
Tensile	32.73 lbs	25110 psi	1084%	ASTM D 882-02
	Min Force	Average Force	Max Force	
Tear Resistance	6.77 lbs	7.21 lbs	7.74 lbs	ASTM D1004-03
Puncture Resistance	16.42 lbs	18.61 lbs	19.51 lbs	FTMS 101C

Biocompatibility Flow Chart for the Selection of Toxicity Tests for 510(k)s

Start

Does the device contact the body directly or indirectly? — No → BIOCOMPATIBILITY REQUIREMENTS MET

Yes

Is the material same as in the marketed device? — Yes → Same manufacturing process? — Yes → Same chemical composition? — Yes → Same body contact? — Yes → Same sterilization method?

No

Same manufacturing process? — No → Acceptable justification or test data?
Same chemical composition? — No → Acceptable justification or test data?
Same body contact? — No → Acceptable justification or test data?
Same sterilization method? — No → Acceptable justification or test data?

Acceptable justification or test data? — No ← ; — Yes →

Is the device material a polymer? — No → Is the material metal, metal alloy or ceramic? — Yes → Does it contain any toxic substances e.g., Pb, N, Cd, Zr? — No →

Is the material metal, metal alloy or ceramic? — No ↓

Does it contain any toxic substances e.g., Pb, N, Cd, Zr? — Yes ↓

Is the device material a polymer? — Yes ↓

Adequate justification provided? — No → ; — Yes →

Consult device specific tax profile for appropriate tests.

No Tax Profile ↓

Consult toxicologist for appropriate tests, if necessary.

Consult modified ISO matrix for suggested tests.

Master File, when referenced, has acceptable tax data applicable to the device. — Yes →

No

Submission contains acceptable tax data and/or justification or risk assessment for not conducting appropriate tests. — Yes → ***Toxicologist concurrence, as necessary.***

No

TOXICOLOGICAL DATA REQUIRED

Toxicologist concurrence, as necessary. — Yes → BIOCOMPATIBILITY REQUIREMENTS MET

the ODE, the CDRH, prior to the initiation of expensive, long-term testing of any new device material to ensure that the proper testing will be conducted. Because an ISO standard is a document that undergoes periodic review and is subject to revision, the ODE will notify manufacturers of any future revisions to ISO-10993 that affect requirements and expectations.

3

Containers

Life does not accommodate you, it shatters you. It is meant to, and it couldn't do it better. Every seed destroys its container or else there would be no fruition.

Florida Scott-Maxwell

Disposable containers form the heart of any comprehensive max-dispo facility. To replace dozens of hard-walled (steel or glass) containers that are used to store media, starting materials, and intermediate and finished products, whether kept at room temperature or kept frozen, there is a great need for containers. Fortunately, disposable bag systems have been very well adopted as alternates to hard-walled containers. And this is because, historically, pharmaceutical products, such as sterile intravenous solutions, blood, plasma, plasma expanders, and hyperalimentation solutions, have been stored and dispensed in these types of bags. For blood storage, a disposable bag would have one-layer films made from polyvinyl chloride (PVC) or ethylene vinyl acetate (EVA).

Given in the following is a listing of major suppliers of disposable containers. Most major equipment suppliers have proprietary bags to fit only their equipment, and while generic bag manufacturers may have alternates to these proprietary bags, there are intellectual property issues involved as many of these bags may have patent protection.

Proprietary Bag Suppliers

Thermo Scientific (www.thermoscientific.com)

Sartorius-Stedim (www.sartorius-stedim.com)

Pall (www.pall.com)

GE (www.gelifesciences.com)

Millipore (www.millipore.com)

Xcellerex (www.xcellerex.com)

LevTech by ATMI Life Sciences (http://www.atmi.com/lifesciences/)

New Brunswick Scientific (www.nbsc.com)

Generic Bag Suppliers

Advanced Scientifics (ASI, www.advancedscientifics.com)

- PL-01077 polyethylene single-use bag is a 5 layer 7 mil co-extrusion film that provides a barrier and durability. Utilized on smaller bag sizes up to 1 L, it maintains comparable, extra values to larger polyethylene (PE) bags.

- PL-01026/PL-01077 polyethylene single-use bags are produced utilizing two films. The fluid contact film is a 5 mil polyethylene (PL-01026). The outer is a 5-layer 7 mil co-extrusion film that provides barrier and durability (PL-01077).

- PL-01028 ethyl vinyl acetate single-use bags are produced utilizing a single film. The film is a 4-layer 12.5 mil co-extrusion film that provides barrier and durability.

- Drums, protective containers, and tank liners

- Containers/fill port automatic aseptic filling: when used in conjunction with good technique and a laminar flow hood, this yields an aseptic bag fill. The semiautomatic filling system utilizes a fixture and cap assembly developed and manufactured by ASI, and fully controls the filling interface with no user interaction required with the fill port. What is left after completion is a tamper-evident dispensing port. This results in a cleaner, more efficient, and cost-effective method of filling.

Charter Medical (www.chartermedical.com)

- Bio-Pak® Cell Culture Bio-Containers are designed for single-use bioprocessing applications, and incorporate Charter Medical's Clear-Pak® film that was chosen by Charter Medical for its superior clarity and excellent performance in promoting cell growth and viability. The Clear-Pak® film is a single-web, multilayer, co-extruded film that provides excellent gas barrier properties to minimize pH shift for greater product stability.

- Bio-Pak® 3D Gusset Bio-Containers are available in a range of sizes from 50 L to 1,000 L. The 3D gusset design is ideal for preparation and storage of media and buffer solutions.

- Bio-Pak® Small Volume Bio-Containers are designed for bioprocessing applications, storage, and transport of sterile fluids. They are available in sizes ranging from 50 mL to 20 L. The boat port design provides flexibility in tubing interface options and facilitates maximum recovery of stored materials.

- Bio-Pak® XL & XLPlus Bio-Containers are an efficient, lightweight, and cost-effective alternative to large tanks and totes for sterile fluid containment and processing. The single-use Bio-Pak® XL bags eliminate the issues surrounding cleaning validation, storage, and sterilization of traditional bio-containers.

- Contour Tank Liners are a cost-effective alternative to dedicated tanks and totes. Contour liners reduce cleaning validation and sterilization of traditional containers. Most importantly, because they are single use, the potential of cross-contamination between different products is reduced.

- Bio-Pak® Totes are application-designed mobile totes mounted on durable, nonmarking wheels. These stainless steel totes hold the flexible bag plus outlet tubing in a self-contained, wheeled unit that can be safely transported by forklifts. A unique bottom outlet system allows fast flow rates and minimal container holdup volume.

- Freeze-Pak™ Cryogenic Bio-Containers are designed for use in cryogenic temperature applications under liquid nitrogen conditions, and are used predominately for clinical and research applications. The Freeze-Pak™ cryogenic film is a single-web polyolefin monolayer of 12 mil thickness and is preferred based on the film's performance during the freeze–thaw process.

Applied Bioprocessing Containers (http://www.appliedbpc.com)

- Small volume containers, 50 mL to 20 L, with integrated handle, integrated hanging capability, and needle-free sampling port, which may be used with a sterile welder and is available as a manifold system.

- Containers for cylindrical tanks, 50 L to 750 L, 2D and 3D designs, top or bottom drain, and available as a liner, fit most cylindrical tanks and is available as a manifold system.

Disposable bags are made from plastic films (Chapter 2), whose composition is determined by the need for robustness, performance, and often the size of the container. These bags have multiple layers for strengthening the walls. Given in the following is the construction of ASI's typical bag design (Figure 3.1):

FIGURE 3.1
Layers of plastics in PL-01077 bag film offered by ASI.

There is a wide choice available from 1 L to 3,500 L bags with a variety of shapes, volume, available ports, tubing, in-line filters, and any other custom feature besides the standard offering by these manufacturers. Generally, it would be advisable to use an off-the-shelf item even though the generic manufacturers offer custom bags readily: the reason for this choice is that there is likely to be a larger volume of data available on off-the-shelf bags, and also they are likely to be available on an as-needed basis.

The typical applications in bioprocessing use tank liners and 2D and 3D bags.

Tank Liners

Tank liners are simple, disposable bags used to line containers and transportation systems. In most cases, they are not gamma sterilized since these are used in open systems most of the time, such as in the preparation of buffer solutions and culture media at the first stage of preparation. The container within which the liner is inserted is there only to provide mechanical support.

Commercially available overhead mixers can readily be integrated because these systems are open. A broad choice of low-density polyethylene liners are available from vendors that supply to several industries reducing the cost of liners. Disposable equipment suppliers also offer these choices. For example, Thermo Scientific's HyClone tank liners are designed for use with commercially available overhead mixers. The chamber is constructed of CX3-9 film with dimensions optimized for Thermo Scientific HyClone standard drums and commonly used industry standard cylindrical tanks. Top-entry standard products for maximum recovery using industry standard connection systems in unit volumes of 50, 100, and 200 L. Tanks are supplied sterile to minimize bioburden. A dolly is available to provide mobility of volumes up to 500 L.

The hard-walled containers are necessary in the preparation of buffers and media as this offers the cheapest alternative; however, these containers do not contact any formulation component and, as a result, the cheapest containers should be used. The most likely choice would be a plastic off-the-shelf drum, such as a 55-gallon drum. Several major equipment suppliers provide a complete line of mixing systems and, while these do offer an advantage in handling large volumes consistently, one can readily put together a system from off-the-shelf components at a substantially lower cost. It is noteworthy that the more expensive systems come with programming elements that might make the Process Analytic Technology (PAT) work easier but, at the stage of buffer and media preparation, the challenges are few and readily overcome by implementing the simplest and cheapest systems. This is what is intended in the max-dispo concept—to use only what adds value.

2D Fluid Containers

For smaller volumes, 2D bags work well, from less than 1 L to 50 L, before they become difficult to handle. The largest 2D fluid container for bioreaction is provided by GE for their Cellbag operations (Wave Bioreactor) in 1,000 L size; other suppliers such as Charter Medical can provide containers up to 3,500 L in size. These bags are produced from two-layer films that are welded together at their ends. The result is a flat chamber that has ports either face welded or end welded. The choice of ports is determined by the user and most suppliers have standard combinations that might work well in most instances. It is important to iterate here that any custom-designed bag or configuration would require new studies to establish the role of leachables; this may not be necessary if standard off-the-shelf items are used that have already gone into cGMP manufacturing and approval of products made using them.

Besides their use as bioreactors, the 2D bags are utilized in a reclining or hanging position as manifolds for sampling, dispensing, and holding the product.

2D Powder Bags

In some instances, it may be necessary to use bags to store powders (such as buffer salts, API, and excipients): these bags have a funnel shape and are equipped with large sanitary fittings or aseptic transfer systems, and are antistatic and free of additives. An example of such a bag is the Thermo Scientific HyClone Powdertrainer. Large-size powder bags are generally custom-designed.

3D Bags

The 2D bags have an interesting problem in their design: at a larger scale it becomes difficult to maintain their integrity. The 1,000 L bag offered by GE is recommended to be used with no more than 500 L of media; beyond that, the seals may not hold since the weight of the fluid inside is transferred to the seams of these bags. This becomes particularly problematic when the 2D bags are rocked or shaken, which adds stress to the seams.

3D bags as liners in hard-walled containers obviate the problems of integrity with 2D bags; today, these bags are available in sizes of 3 L to 4,000 L sizes. The 3D design also provides additional surface to install ports with complex functions and at both top and bottom. The 3D bags are made by welding films and are mostly offered in cylindrical, conical, or cubical shapes. Often, the shape is determined by the method of how these containers are stored or stacked in outer containers that have the same shape, which allows a snug fitting of the 3D bags. While a very large liner can always be brought into the manufacturing area, the outer containers are at times built before the facility is completed; companies offering modular construction of outer containers would do well in the future if they offer an option of assembling an outer container from smaller pieces.

To facilitate their use such as in buffer preparation, these outer containers may be equipped with weight sensors, recirculation/mixing fluid management, and temperature control if required. The temperature control can be achieved in several ways, the cheapest one being wrapping them in blankets that are temperature controlled, and the most expensive being to use jacketed containers with circulating fluids. The weight measurement is of greatest importance and, while most manufacturers would use a floor scale, large-scale production requires installation of load cells in the outer containers to avoid moving the containers for weighing.

Transportation Container

Products at different stages of manufacturing often need to be transported within the company or to remote locations to complete the process; finished products are also shipped out to customers and this requires the selection of safe, stable, and closed container systems that maintain sterility. Examples of these containers include

Flexboy, Flexel 3D Palletank, and Celsius FFT products (www.sartorius-stedim.com)

Nalgene (www.nalgene.com)

Thermo Scientific (www.thermoscientific)

BioShell™ container system designed to protect single-use bags during storage, handling, and shipping. High-purity, dual-density foam construction can withstand multiple impacts at −70°C (www.bio-shell.com)

Disposable bags can be readily used to transport or store frozen products, from cell culture as Working Class Bank (WCB) for direct introduction into a bioreactor to shipping biological API; while flexible bags can survive temperature variations, often it is difficult to detect damage to them during transportation and, thus, they require a protective surface around them to obviate this risk.

Summary

- Plastic disposable containers offer the best solution in disposable components utilization as they remove the cleaning and validation requirements.

- Low-density PE liners in a hard-walled plastic container and a standard mixer make the cheapest combination of pieces to prepare buffers and media.

- More complex mixing systems are not necessary, and neither are the expensive proprietary containers to hold these PE liners.

- 2D bags can be used only for smaller-size volumes, while 3D bags with an outer nondisposable container increase the limit of fluids that can be contained to thousands of liters.

- Several novel shapes and sizes are available to fit just about any need.

- Flexible bags can be used for the transportation of biological drugs and, while they survive freeze–thaw cycles, it is often difficult to record breaches in their integrity, thus requiring an outer protective surface.

- Custom-designed bags are readily available, but these are very expensive and do not give the user the benefit of the large database provided by the vendors, so one should stick to off-the-shelf products whenever possible.

- Future novel uses of bags may include storage of WCB for direct addition to bioreactors.

4

Mixing Systems

The unit operation of mixing is extensively involved in bioprocessing systems. Some of the keys to mixing operations include mixing to dissolve components of a buffer, culture media, refolding solution, dispersion of cell culture in bioreactors, and heating or cooling of liquids.

All mixing operations must be fully validated as part of PAT to ensure that optimal mixing has been achieved all the time. While the stainless steel mixing vessels have long been used and the principles behind mixing and demixing of components with traditional mixing devices have long been studied, much remains to be understood about achieving homogenous mixtures in disposable bags.

In bioprocessing operations, two types of mixing are important: one that leads to the dissolution of solutes, and the other that provides a homogenous environment such as in a bioreactor or a refolding tank. How fast a mixture of powdered components in a buffer mixture dissolves will depend to a great degree on the solubility of individual components, the agitation applied, and the temperature and length of mixing. In theory, mixing involves distributive, dispersive (breaking of aggregates), or diffusive steps. All of these steps require energy that is provided by the mechanical motion induced in liquids. A laminar movement of liquid or a turbulent movement can achieve the mixing, and the Reynolds number (Re) of mixing obtained can predict this.

In fluid mechanics, the Reynolds number is a dimensionless number that gives a measure of the ratio of inertial forces to viscous forces and consequently quantifies the relative importance of these two types of forces for given flow conditions. Laminar flow occurs at low Reynolds numbers, where viscous forces are dominant, and is characterized by smooth, constant fluid motion; while turbulent flow occurs at high Reynolds numbers and is dominated by inertial forces, which tend to produce chaotic eddies, vortices, and other flow instabilities. In a cylindrical vessel stirred by a central rotating paddle, turbine, or propeller, the characteristic dimension is the diameter of the agitator (D). The velocity is ND (where N is the rotational speed (revolutions per second)), μ is the kinematic viscosity, and ρ is the density of fluid. Then the Reynolds number is

$$Re = \frac{\rho ND^2}{\mu}$$

The system is fully turbulent for Re values above 10,000.

In fluid dynamics, mixing length theory is a method attempting to describe momentum transfer by turbulent Reynolds stresses within a fluid boundary layer by means of an eddy viscosity. The mixing length is the distance that a fluid parcel will keep its original characteristics before dispersing them into the surrounding fluid.

Laminar mixing, often encountered in fluids with high viscosities, originates from a longitudinal mixing where fluid motion is dominated by linear viscous forces. Fluid particles flow along parallel streamlines and, to obtain homogeneity, radial mixing is necessary, which can be achieved through mechanical forces such as using a stirring bar or an impeller or rocking the base. Thus, turbulent mixing provides the greatest effectiveness as evidenced by the utility of high-speed stirrers.

Manufacturing processes are validated for their outcome in a cGMP environment. As a result, the desired mixing quality, which in most cases is a homogenous mixture, is obtained by mixing for a certain period of time (with a range) and with a certain force applied (such as rpm, rocking motions per minute, or other such parameters) and, in those instances where a demixing may occur, a time for which the mixture remains homogenous. Generally, for most of the mixing processes encountered in bioprocessing, these parameters are easy to study and validate. The most difficult one is the mixing of culture in a bioreactor, a topic that will receive greater discussion in the next chapter.

Types of Mixing

There are several distinct types of mixing systems currently available in bioprocessing where disposable mixing containers are used. These include

1. Stirrer systems
   a. Rotating stirrer
   b. Tumbling stirrer
2. Oscillating systems
   a. Rocker
   b. Vibrating disc
   c. Orbital shaker
   d. Pedal push
3. Peristaltic system
   a. Recirculating pump

The systems using stirrers can have the stirring element either driven magnetically or connected through a sealed shaft. Oscillating types mix by moving the liquid inside a bag (mostly 2D types) by rocking them or shaking using mechanical vibrations or ultrasonic vibrations. Generally, the mixing systems that do not involve any mechanical parts inside the bag (either 2D or 3D) are preferred to reduce the cost, the risk of damage to the bag from rotating devices, the grinding of bag, or the stirrer inside the bag; those stirring systems that use a magnetic field provide better sterility compared to those that are magnetically coupled.

Stirring Magnetic Mixer

- XDM (Xcellerex), 100 L to 1,000 L, the XDM Quad Mixing System comprises an integrated magnetic stirrer with a compact motor, a bottom-mounted disposable stirrer; the coupling between the motor and the disposable stirrer is magnetic. The square configuration offers enhanced mixing efficiency through a natural baffling effect and compact storage capability. The bottom is slanted to ensure a low residual volume after discharge.

- The Flexel 3D LevMix System for Palletank, 50 L to 1,000 L, combines the LevTech levitated impeller licensed by ATMI and the Sartorius-Stedim Flexel 3D Bag. It comprises a stainless steel, cube-shaped container with a door for ease of bag mounting. In addition, it has windows to enable observation of the mixing process, a drive unit for levitating or rotating the stirrer, and a disposable bag with a center-mounted magnetic stirrer.

- Magnetic Mixer (ATMI Life Sciences), 30 L to 2,000 L.

- Jet-Drive (ATMI Life Sciences), 50 L to 200 L.

- Mobius (Millipore), 100 L to 500 L.

- LevMixer (ATMI Life Sciences), 30 L to 2,000 L, is ultraclean as it does not produce any residue from mechanical motion, suitable for downstream operations as well.

Stirring Mechanical Coupling Mixer

- S.U.M. (Thermo Fisher Scientific), 50 L to 2,000 L; there are two types of magnetic stirrers driven by a stirring plate available for different mixing applications. Not intended for sterile applications: suitable applications include dissolving solid media and/or buffer components prior to sterile filtration.

- Thermo Fisher Scientific HyClone Mixtainer Systems with an impeller linked to an overhead drive and is coupled by a sealed bearing assembly, which maintains the integrity of the system. The mixing stirrer is installed off-center. This mixer is intended for powder–liquid and liquid–liquid mixing and has sterile disposable contact surfaces.

Tumbling Mixer

- Pad-Drive (ATMI Life Sciences), 25 L to 1,000 L, uses a tumbling stirrer mounted from top; the wand rotates inside an inert polymer sleeve.
- WandMixer (ATMI Life Sciences), 5 L to 200 L, uses a tumbling stirrer whose axle is built into the bag from the top of the bag.

Oscillating Mixer

- Wave (GE Healthcare), 20 L to 1,000 L, horizontal oscillation on a rocker. The rocking motion is very efficient in generating waves, and the wave-induced motion in the bag causes large volumes of fluid to move facilitating the dispersion of solids. The optimum operating parameters depend on the combination of the container geometry, bag support, filling volume, rocking angle, rocking rate, and the characteristics of the mixture (solids, foam, etc.).
- HyNetics (HyNetics Corporation), 30 L to 5,000 L, vertical oscillation of a disk or septum. The key feature is the mixing disk, which is fabricated from rigid, engineered polymers. Multiple, evenly distributed slots penetrate the disk. The underside of the disk incorporates pie-shaped flaps. These flaps open as the disk moves up from the bottom of the mixing bag on the drive's upstroke, allowing fluid to flow through the disk's slots. The flaps close on the downstroke, forcing the liquid toward the bottom of the vessel and subsequently up the walls of the vessel. The mixing disk, flaps, polymer mixing shaft, and the shaft rolling diaphragm seal, which attaches to the bag film, are disposable.

- SALTUS (Meissner), 5 L to 2,000 L, vertical oscillation of a disk or septum; based on a vibrating disk with conical orifices. Due to the oscillating movement and the conical orifices, liquid jets develop at the tapered end of the holes. Thus, an axial fluid flow pattern is achieved. The frequency and amplitude of the vibration can be adjusted to provide either vigorous or gentle mixing. The bag is pre-assembled with the rigid high-density polyethylene (HDPE) vibrating disks and with tubes, filters, and a sampling port, in addition to a disposable sensor plate for pH, dissolved oxygen (DO), and temperature measurement. Due to the frictionless, oscillating movement of the disk, it can be used where an ultraclean environment is required.
- PedalMixer (MayaBio, www.mayabio.com), 10 L to unlimited volume, no stirring device, uses a pedal outside of the bag to push the liquid to mix, and can be used with any generic bag; this is the lowest-cost option. The newest type of oscillating system is a patented pedal system whereby the 2D bag remains stationary on a flat surface and a pedal pushes at one end of the bag creating waves inside the bag; a slight tilt of the platform imparts potential energy to the contents, while the kinetic energy moves the liquid and provides a mixing profile identical to that obtained using a rocking platform or any other form of the use of mechanical energy. A significant advantage of this system is that it can accommodate any size (since no stress is produced on the bag) and can accommodate all shapes and sizes of bags, allowing the use of generic bags. (See Figure 4.1)

Peristaltic Mixer

- The Flexel3D Palletank for recirculation mixing incorporates one or two recirculation loops and can be equipped with Sartorius-Stedim's Mechatronics load cells to facilitate fluid management.

Summary

- A large number of unit operations in bioprocessing involve mixing; fortunately, these are relatively simple operations that are easily validated.
- The largest mixing operations involve buffer and media preparation that can involve thousands of liters. Since these components are

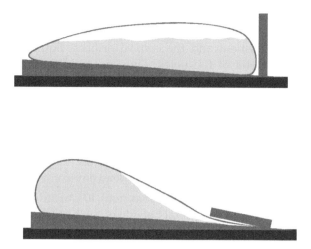

FIGURE 4.1
MayaBio PedalMixer mixing system.

sterilized, likely by filtration, it is not necessary to use any special proprietary mixing system. An off-the-shelf plastic drum with a PE liner and industry-standard mixers can do the job well at a fraction of the cost. It is not necessary to use any proprietary liners as long as the user is able to qualify a supplier; at this stage, the qualification is relatively simple. Since all of unit operations in a cGMP operation are validated, once a system has been qualified, it can be used repeatedly.

- Open mixing of media and buffer may be provided with a laminar hood in those environments where there is a risk of cross-contamination to reduce any additional burden on filter systems.

- The mixing systems available today are the same as used in disposable bioreactors: in some instances the platform can be used for both operations.

- While many reputable suppliers have developed highly sophisticated 3D systems, these are not necessary for buffer and media preparations; the cost of 3D bags with built-in stirring systems can be prohibitive.

- The 2D bags offer many advantages including the ease of storage because they are horizontally expanded; the wave motion created inside these bags is extremely efficient.

- The newest entry in mixing technology, which is from MayaBio, makes it possible for users to use any generic bag for mixing, further reducing the cost as well as reducing the dependence on a proprietary supplier of components.

- The power requirements in operating the mixing systems are the lowest in nonstirring types, such as the oscillating mixers; however, this is not a major consideration in the overall cost of mixing.
- In the future, several novel systems and the utilization of existing systems will appear in the market, and there are likely to be greater integration of the various steps of bioprocessing.

5

Disposable Bioreactors

Piety is the fermentation of the forming mind and the putrefaction of the disintegrating one.

Franz Grillparzer (1791–1872), Austrian author
Notebooks and Diaries **(1838)**

Hard-walled bioreactors have been used for centuries, from kitchen utensils to multistory stainless steel behemoths; the field of bioreactor design has remained pretty much the same for a long time. The essential elements of a bioreactor—a utensil to contain a culture and media with sufficient mixing and aeration—are readily provided in the traditional designs of bioreactors. Today, we have a multitude of options in the design of bioreactors, and these came about once the use of bioreactors expanded to the manufacture of biological drugs requiring many control features that were not needed or required in other industries. With the use of animal, human, and plant cells and viruses to produce therapeutic proteins, vaccines, antibodies, etc., there arose a need to modify the traditional bioreactors to accommodate the growth needs of these new production engines: recombinant engineering put these new engines in the forefront of biological drug production. One major change in the design of bioreactors that is recent is the use of disposable bioreactors to avoid the challenges of cleaning validation, thus reducing the regulatory barriers in drug production. Hundreds of new molecules are under development using disposable bioreactors, and in many instances disposable bioreactors are used to manufacture clinical supplies. Yet, no drug has been approved for marketing that is manufactured on a commercial scale using a disposable bioreactor. However, this situation will soon change as the new molecules under development move further in the approval cycle.

Almost all of recombinant drugs in the market today were developed by large pharmaceutical companies starting about 30 years ago when the only choice available was the traditional bioreactor; even though their process may be less efficient, it is not worth the effort to switch over to another manufacturing method because of the prohibitive cost of changeover protocols that need to be completed. A case in point is the use of roller bottles to manufacture erythropoietin: Amgen, the world's largest producer of erythropoietin, continues to use roller bottles despite their inefficiencies and risks, but for new products Amgen will be using stirred bioreactors.

Disposable bioreactors have varied designs and purposes but all of them are made of Class VI plastic films, are sterilized by gamma radiation, and are disposed of after use; they may come with several attachments that allow the filtration of media and monitoring of pH, DO, OD, pCO_2, temperature, and other PAT-related parameters. Use of stirrers and paddles, and shaking and rocking the bags by mechanical or hydraulic means achieve mixing and aeration inside the bag. A choice of aeration systems may include surface aeration (e.g., in Wave Bioreactors) to forced sparging through proprietary ceramic tubes (e.g., MayaBioReactors). The host cell yields obtained using disposable bioreactors match or exceed those obtained in traditional reactors.

Disposable bioreactors come in many sizes, from milliliters to thousands of liters; they can be equipped with bioinformatics systems that range from very simple to very complex; they can be manual or highly automated; they can be as inexpensive as a plastic bag to as expensive as the high-end traditional hard-walled bioreactors. The disposable bioreactor industry is still evolving, with new inventions surfacing almost routinely. Here is a brief look at their historical development over the past 60 years:

First Period—First Ten Years (1960s): Petri dishes, T-flasks, roller bottles, shaken plastic bags. At first, the glass petri dishes were replaced by plastic plates, and the most significant development was the use of polypropylene and Teflon bags by the Krolinska Institute in Sweden to grow bacteria and yeast cells, albeit on a very small scale.

Second Period—Next Thirty Years (1970s to 1990s): Disposable hollow fiber system, two-compartment system, multitry cell culture, static bags for cell expansion, pneumatic mixing (peristaltic recirculation), and rocking bags. The hollow fiber technology required recirculation of media to grow anchored or suspended animal cells using the Cellmax HFBS (FiberCell), the AcuSyst-HFBSs (BioVest), and the Xcell HFBS (BioVest). These bioreactors were able to operate continuously for months at a time and helped produce quantities ranging from a subgram to a few grams. Even though high cell densities could be achieved, the problems of scaling up these bioreactors ruled them out as a viable option for commercial production, and they are used today to make small quantities of test substances.

The Cell Factory made of polystyrene was a flask-like culture system containing a number of trays stacked in parallel in a single unit. This was a good scale-up model for commercial production and replaced roller bottles used for adherent cells.

CellCube from Corning Costar is similar to the Cell Factory, runs in perfusion mode, and proves useful for adherent cell lines; it was used for vaccine production on a limited scale and never showed potential for commercial therapeutic protein manufacturing.

The LifeReactor was a peristaltic pump-driven bubble column biore-actor, where mass and heat transfer are achieved by direct sparging of a conical-shaped disposable culture bag (1 L to 5 L). It was mainly used for the growth of plant origin organ cultures.

The two-compartment dialysis membrane bioreactors have a semi-permeable membrane that separates the cells from the bulk of the medium and again permits continuous diffusion of nutrients into the cell compartment with simultaneous removal of waste products. The two models, the MiniPerm (Greiner Bio One) and the T-flask-based CELLine (INTEGRA Biosciences, Sartorius-Stedim), must be kept in a CO_2 incubator, and achieve high cell density allowing antibody production.

Third Period—Next Twenty Years (1990s to current): Wave-mixed reactors, stirred 3D reactors, orbitally shaken reactors; used in clinical sample production, small-scale commercial manufacturing, and comprehensive disposable systems. While the use of a shaken bag goes back to the 1960s, it was not until around 1996 when Vijay Singh disclosed his invention of the Wave Bioreactor (Figure 5.1) and marketed them in 1998 that the industry woke up to a new reality in biological drug manufacturing. While the original Wave Bioreactors served the purpose well, soon it was realized that many of the shortcomings, such as inability to grow bacteria or scale up to larger volumes, were overcome recently (2010) by the finding of MayaBioReactors (www.mayabio.com) that all types of cells can be grown in 2D bags.

FIGURE 5.1
Wave bioreactor.

FIGURE 5.2
Comparison of various stirring systems in 3D disposable bioreactors.

Given below is a description of the various methods used to induce motion inside a disposable bag:

Types of stirring mechanisms (Figure 5.2). From left, stirrer mechanically attached to a motor; stirrer magnetically levitating, no contact with motor; magnetic stirrer at bottom, rubs off the surface, mechanical stirrer inserted from top.

Rocking wave motion is the most commonly used; pioneered by the Wave Bioreactor, several equipment suppliers have adopted this system (Figure 5.3).

Stationary bioreactor concept differs significantly from the usual wave motion that requires moving the base of plate; here, the bag stays stationary and a flapper instead pushes down one edge of the disposable bag.

Somehow the concept of using 2D bags to grow host cells did not pan out widely, and most of the major equipment leaders, such as Sartorius-Stedim, Pall, EMD Millipore, New Brunswick Scientific, and Thompson Scientific, adopted 3D versions of disposable bioreactors. The recent entrant to the race is Xcellerex, which has done well with its large-scale 3D bioreactors. The success of Xcellerex comes from its reputable customer support as they build out the equipment as client solutions, while others position themselves as equipment suppliers.

There is no doubt that the simplest and the most cost-effective bioreactors are the 2D or pillow types as they do not require an outer container and by design avoid any internal stirring. The wave-mixed bag systems represent one of the largest groups among single-use bioreactors and include the

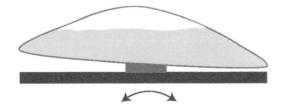

FIGURE 5.3
Liquid motion in wave-based mixing systems.

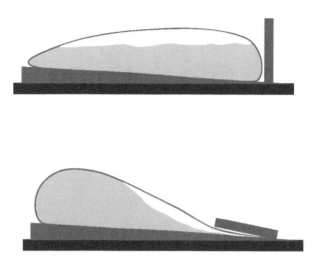

FIGURE 5.4
New concept of mixing in MayaBioReaxtors by pushing the liquid against an incline.

AppliFlex, the BIOSTAT CultiBag Rocking Motion, the BioWave, the CELL-tainer, the Wave Bioreactor, the Wave and Undertow Bioreactor (WUB), and the most recently introduced MayaBioReactors, which have a stationary surface and require the least amount of energy input (Figure 5.4).

Perhaps the equipment suppliers' profit margins were not large enough or perhaps they understood the psychology of the industry well enough to know that it would not be easy for Big Pharma to come down from towering bioreactors to lay-flat bags with rocking motion as the manufacturing equipment. This caused the proliferation of 3D technologies. The stirring bag bioreactors were first introduced by Thermo Fisher's Single-Use Bioreactor (SUB), developed as a result of cooperation between Baxter and Hyclone and currently the market leader; this and the XDR-Disposable Stirred Tank Bioreactor from Xcellerex were the only such systems available initially. This was followed by the Nucleo bioreactor (ATMI Life Sciences), the BIOSTAT CultiBag Stirred (STR) (Sartorius Stedim) (Figure 5.5), the Mobius CellReady bioreactor (EMD Millipore/Applikon) (Figure 5.6), and the CelliGEN BLU Single-Use Bioreactor.

Xcellerex Bioreactor

The nonstirring type 3D reactors include Sartorius-Stedim's SuperSpinner D 1000, which is noninstrumented; its aeration comes from hollow fiber mem-

FIGURE 5.5
Cultibag bioreactor.

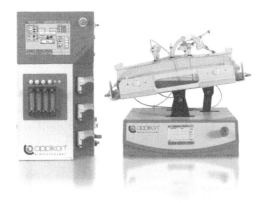

FIGURE 5.6
Applikon bioreactor.

branes wound around a tumbling stirrer, all features making it a simple reactor to operate and test the expression of new cell lines (Figure 5.7).

Orbital shaker bioreactors comprise the third-largest category after the wave and stirring types and promise high-throughput systems for scaling up to pilot scale. Screening systems such as the M24 Microbioreactor (Applikon, Pall Life Sciences), the BioLector (mp2-labs), and the Sensolux (Sartorius-Stedim) are typically equipped with noninvasive single-use sensors useful for PAT work. Sine orbital shaking was first applied to flasks and plates, and these can still be upgraded to reactor level by connecting them to a PreSens's Sensor Dish Reader (SDR) or a Shake Flask Reader (SFR) using precalibrated sensor patches for pH and DO. The CultiFlask 50 disposable bioreactor and the Disposable Shaken Bioreactor System (a cooperation between ExcellGene, Kiihner, and Sartorius-Stedim) and the CURRENT Bioreactor (AmProteins) serve as midsize reactors, Zeta's bio-t

FIGURE 5.7
Xcellerex bioreactor.

(a proof-of-concept reactor) is a bag bioreactor with a Vibromixer, where the movement of a perforated disk fixed on a vertically oscillating hollow shaft causes an axial flow in the bag, which mixes and aerates the cells. The form, size, and position of the conical drill holes on the disk affect the fluid flow and oxygen transfer efficiency in the bag and contribute to the elimination of vortex formation. Similarly, the BayShake Bioreactor achieves vertical oscillation in which the culture broth oscillates in a surface-aerated cube-shaped bag.

The bubble bioreactors are exemplified by Nestle's Slug Bubble Bioreactor (SBB), which generates intermittent large, long, bullet-shaped bubbles termed "slug bubbles" that occupy nearly the entire cross section of a tube, are generated at the bottom of the bag, and rise to the top. To provide a determined quantity of air at a given frequency, a solenoid valve is used to control bubble generation. Varying the air inlet pressure and the valve opening time controls the quantity of air and the bubble frequency.

The Bioreactor System (PBS) works with an air-wheel design and a dual sparger system for efficient mixing and aeration. In the case of the CellMaker systems (Cellexus Biosystems), the unique asymmetric shape of the culture bag is significant. The CellMaker Regular is a single-use bubble column. This system is preferable for microbial productions. The version specific to animal cell cultivations is the hybrid CellMaker Plus, where pneumatic and mechanic drives are combined. Mixing and aeration is achieved by transverse liquid movement. While the airflow is induced by a sparger tube, the two magnetically driven propellers intensify the "riser" flow. Excessive foam formation, which is linked to flotation and is a well-known problem in bubble columns, may be minimized or even eliminated by applying pressure to the headspace within the bag.

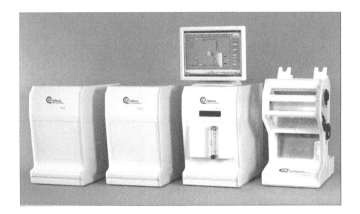

FIGURE 5.8
Celluxus bioreactor.

Cellexus Bioreactor

With exception of the WUB, the SBB, and the microbial versions of the XDR, CELL-tainer and CellMaker, all disposable bioreactors have been developed primarily for fed-batch operations with animal suspension cells (Figure 5.8). This kind of operation is most common in biomanufacturing. Anchorage-dependent (adherent) cells are less widespread in today's processes; however, disposable bioreactors such as AmProtein's CURRENT Perfusion Bioreactor do allow the cultivation of adherent cells if they are grown on microcarriers. Microcarriers also support the cell attachment to a 3D structure, enabling a higher cell density and productivity, and culture conditions that are nearly identical to an in vivo environment.

CELL-Tainer Cell Culture System

The fixed-bed bioreactors include the FibraStage (using FibraCel disks in four disposable bottles per bioreactor system, with a maximum volume of O.5 L CV per bottle) from New Brunswick Scientific and Artelis's fixed-bed bioreactor (iCELLis bioreactor, with a maximum volume of 500 mL per packed bed). Both bioreactors, which require microcarriers, were specifically designed for the production of cell-culture-based animal cells. The FibraStage is kept in an incubator and is suitable for production at a laboratory scale (Figure 5.9).

A novel small 3D bioreactor is Hamilton's BioLevitator operating with modified, surface-aerated 50 mL plastic tubes, which oscillate vertically. The

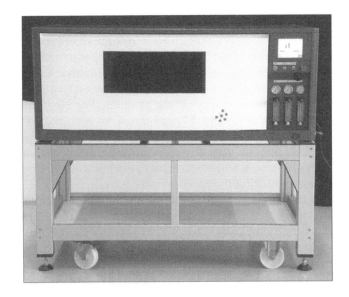

FIGURE 5.9
CELL-tainer culture system.

fully automated SimCell MicroBioreactor System (with parallel disposable cassettes and six microbioreactors per plate) ensures efficient process optimizations for animal cell cultures, which can be transferred to stirred processes with high reproducibility.

Wave-Mixed Bioreactors

These comprise a bag that consists of a multilayer film; ethylene vinyl acetate (EVA) is the contact layer in most cases. The mixing takes place in the bag by moving the platform sections. Oxygen is taken up from surface renewal of waves formed, leaving a bubble-free surface.

A variety of designs, degree of bioinformatics, and sizes are available in this category. Except for the WUB and CELL-tainer, the wave is caused by a one-dimensional horizontal oscillation of the culture broth in the bag located on a rocker unit. The intensity of the mass and energy transfer and, therefore, cell growth and product expression can be directly controlled through wave generation and propagation. These features are adjustable by modifying the rocking rate, the rocking angle, the filling level of the bag (up to 50% maximum), and the aeration rate of the Wave Bioreactor, the BioWave, its successor (the BIOSTAT CultiBag RM), the AppliFlex, the Tsunami Bioreactor, and the CELL-tainer.

In the Tsunami Bioreactor, up to six rocker units integrated into one rack housing 5 bags (each with 160 Liters Culture Volume (L CV) or 64 bags (each with 5 L) move in opposite directions. This is no longer available.

Oxygen transfer (which is described by the volumetric oxygen transfer efficiency rate (K_La) values) and its influence on the cultivation result have been investigated for the majority of the systems. For Newtonian culture broths, K_La values between 5 and 30 per hour were reported as being typical for animal cell cultivations in the BioWave, the Wave, and the AppliFlex. Oxygen limitations may be virtually disregarded during such a process as increasing the rocking rate, and angle is more effective in increasing the oxygen transfer than increasing the aeration rate.

The required high oxygen level can be achieved by operating a BIOSTAT CultiBag RM with low CV (50 L bag with 5 L CV) or the CELL-tainer. In contrast to the version for cell cultures (CELL-tainer Bioreactor) where K_La values exceed 100 per hour, values above 200 per hour are possible in the version for microbial cultures (CELL-tainer Microbial Bioreactor). This is attributed to the 2D movement of the CELL-tainer, which ensures higher oxygen transfer rates for microorganisms.

In the WUB, the wave propagated inside the bag is generated by periodic upward movement of the movable head and/or foot section of the horizontal table (platform) on which the bag is located; the K_La values of the WUB are similar to those achieved with the BioWave. The parameters having the most impact on the K_La data are the angle of the platform, the percentage of the CV located on and lifted by the platform, the aeration rate, and the time taken for the platform to complete one oscillation.

The Wave Bioreactor, BioWave, and BIOSTAT CultiBag RM differ in their sensors and control units.

The wave-mixed bag bioreactors have secured a solid position in mammalian cell-derived seed train manufacturing and process developments aimed at producing therapeutic proteins. These bioreactors are run in a batch, fed-batch (feeding processes), or perfusion mode and are preferred reactors for transient transfections; they are becoming widely used in simple, medium-volume processes such as the production of viruses for gene therapies (e.g., recombinant adeno-associated virus vectors) and veterinary as well as human vaccines (e.g., Aujeszky's disease virus, porcine influenza virus, porcine parvovirus, mink enteritis virus, smallpox virus). Traditional disposable virus production bioreactors (roller flasks, Cell Factories) have been successfully replaced by wave-mixed bag bioreactors.

To date, wave-mixed bag bioreactors have proved acceptable for the cultivation of plant cell and tissue cultures in research and development (R&D). Focusing on biomanufacturing, secondary metabolite productions (taxanes, harpagosides, hyoscya-mine, alliin, ginsenosides, isoflavones) have been realized in the BioWave and the WUB. Suspension cells, embryogenic cells, and hairy roots were grown. In addition, the first proteins (e.g., human collagen/alpha, tumor-specific human antibody) were successfully produced with fast-growing suspension cells in the BioWave, the AppliFlex, and the WUB. However, up to now, wave-mixed bag bioreactors have not achieved the same importance for the production of plant cell culture-derived products as they have for animal-cell-based target molecules. The same is the case for microbial products with pharmaceutical significance.

However, a recent modification of 2D bags by MayaBio has made it possible to use the wave-mixed systems for every type of cell and organism; studies reported by MayaBio show bacterial ODs of 70–80 in overnight cultures. The MayaBioReactor introduces a proprietary sparger in Wave Bioreactor bags that allows extensive aeration. Another major difference comes in the platform, which is kept stationary and a pedal pushing up and down at one end of the bag creates wave motions inside the bag, allowing mixing achieved by using a rocking platform.

There is still interest in developing photo bioreactors, and Applikon has recently made a disposable offering (Figure 5.10). A number of recent studies have demonstrated that normal plants could be grown under light-emitting diode light sources very efficiently. These solid-state lamps (SSLs) are tiny semiconductor chips that generate light when powered. The elements that the diode is made from determine the light spectrum it emits. These solid-state devices have been improved over the years and now have greatly increased light intensity and specific wavelengths. These developments have resulted

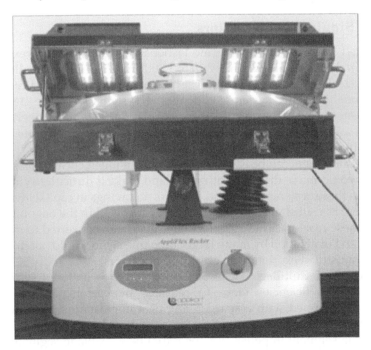

FIGURE 5.10
Applikon photo bioreactor.

in SSL as a self-contained light source for plant growth. Applikon has chosen to develop light panels that are add-on modules for our standard stirred tank and single-use bioreactors. This offers maximum flexibility and a very economical setup for cultivating photosynthesizing organisms. The volume range covers 3 L up to 20 L autoclavable stirred-tank bioreactors and 10 up to 50 L single use bioreactors. SSL plant light has unique characteristics that are useful for plant growth applications. An important characteristic is the spectral distribution of light in the wavelengths region of 450–500 nm and 630–700 nm; these bands are critical for normal plant growth as they fall within the photosynthetically active radiation, PAR, (400–700 nm), which plants primarily use for biological processes and are also favorable for confined applications such as micropropagation.

Another useful characteristic is the long useful life of about 50,000 hours and the high energy conversion efficiency. This results in substantially cooler systems than other light sources. Systems also save energy by using less ventilation and requiring less cooling for growing plants in the culture room. Second, this provides new opportunities for enhancing growth of several hard-to-grow plants or plants that require a specific range of light spectrum.

Stirred Single-Use Bioreactors

The stirred systems sold by Thermo Fisher Scientific and Xcellerex for use with animal cells and for volumes up to 1,000 and 2,000 L offer challenges to stainless steel bioreactors. Both of these bioreactors borrow their dimensions, proportions, sparging systems, and mixing systems from the traditional stainless steel systems. In reality, these are standard stainless steel systems in which a liner has been installed. These bioreactors demonstrate that the way to attract Big Pharma is to offer expensive big machines. For example, the outer containers can be easily replaced with a much cheaper plastic shell but that would make them less attractive and make it difficult to charge the high price these systems command. There is no savings in capital investment while there is a substantially higher expense involved in the ongoing cost to operate these reactors. These reactors have also been converted to microbial versions and evaluation by many large companies who would not mind paying the unjustified exorbitant prices of these reactors.

The BIOSTAT CultiBag STR of Sartorius-Stedim is a closed system and demonstrates efficiencies close to reusable bioreactors. As an option, the bag is equipped with a sparger ring or a microsparger and two axial flow three-blade-segment impellers or a combination of one axial flow three-blade-segment impeller and one radial flow six-blade-segment impeller. Homogeneous mixing in the bag is achieved by the centered stirring system.

ATMI Life Science's Nucleo single-use bioreactors have a cube-shaped bag instead of a cylindrical bag, a tumbling (Pad-Drive) mixing system instead of a rotating impeller, and a dynamic sparging arrangement in place of a static structure and is available in 50 L and 1,000 L volumes (ATMI http://www.atmi-lifesciences.com).

Integrity™ PadReactor™

The Integrity™ PadReactor™ system is a single-use bioreactor specifically designed to fulfill the needs of cell culturists. It is perfectly suited to laboratory environments, process development centers, clinical material supply, and flexible GMP manufacturing. The PadReactor offers an open architecture controller platform, which gives the end-user the opportunity to choose a preferred controller or use an existing control system.

The bioreactor vessel, which offers comparable functionality to classical stirred-tank bioreactors, is a single-use bag integrating an internal paddle mixing and sparger system. This innovative bag design allows a noninvasive connection to the system. The paddle is enclosed in a medical-grade ultra low density polyethylene (ULDPE) sleeve made from the same contact material as the bag itself, and is coupled on top of the vessel with the mechanical mixing head.

As with all ATMI LifeSciences' single-use systems, the Integrity PadReactor utilizes disposable mixing bags made from TK8 bioprocess film. The product-contacting layer of TK8 film is blow-extruded in-house by ATMI under cleanroom conditions using medical-grade ultralow-density polyethylene resin. It is then laminated to create a gas barrier film of exceptional cleanliness, strength, and clarity that is animal-derived component free (ADCF) and complies fully with USP Class VI requirements.

The Integrity PadReactor single-use bioreactor consists of the following:

Drive unit: The flexible drive unit allows the system to cultivate cells in disposable bags. One drive unit can allow the user to mix in multiple disposable mixing bags of various sizes. Each system comes with the appropriate mixing stick for your container.

Mobile retaining tank: The purpose of the retaining tank is to support the mixing bag and provide mobility before and after the operation. Various tank sizes and options are available.

Bioreactor vessel: The reactor vessel uses an innovative bag design that allows a noninvasive connection to the mixer. Mixing is achieved when the integrated paddle/sparger inside the bag rotates within the bag.

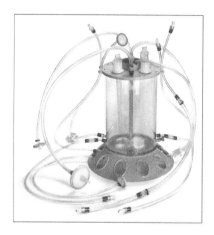

FIGURE 5.11
Mobius CellReady reactor.

Highlights and benefits • Superior mixing capabilities with highly reduced shear stress • Innovative sparging device with better oxygenation and K_La • Adapted for cultivation of suspended or adherent cells at very high densities • Compatible with most cell culture processes • Scalable customizable system • No need for CIP/SIP (disposable bag technology) • Avoid cross-contamination risks • Very low working volume.

The Mobius CellReady 3 L bioreactor is equipped with a marine impeller (top driven), a microsparger or open-pipe sparger, standard sensors, and an Applikon ez-Control process control unit. Similar cell densities and antibody titers can be achieved in the Mobius CellReady, as in stirred 3 L glass bioreactors. A comparable approach to the Mobius CellReady represents New Brunswick Scientific's CelliGEN BLU single-use stirred-tank bioreactor (Figure 5.11).

CellReady Bioreactor

Orbitally Shaken Single-Use Bioreactors

Orbitally shaken bioreactors are very difficult to study because of the free movement of surfaces in the bioreactors. The surface-aerated CultiFlask 50 disposable bioreactor, a noninstrumented 50 mL centrifuge tube with a ventilated cap, can deliver K_La values of between 5 and 30 per hour at CVs of 10 to 20 mL and agitation speeds between 180 and 220 rpm.

Systems with cylindrical bags include the Disposable Shaken Bioreactor System and the CURRENT Bioreactor. AmProtein utlilizes EVA plastic bags in their CURRENT Bioreactor series. It was possible to demonstrate that the oxygen supply (critical for yield optimization) could be improved by the material of construction of the cultivation container in single-use bioreactors.

Bioreactor Selection

Factors to consider include

1. Goal of production, biomass or cell production
2. Bioinformatic controls
3. Scale
4. Biosafety
5. Familiarity
6. Cost
7. Support

More choices are available for animal cell bioreactors. For all kinds of cell expansions and processes based on insect cells, wave-mixed bag bioreactors should be the design of choice. This is especially important if the culture medium used is serum free or protein free (i.e., it contains hydrolysates such as peptones from plants and yeasts), but not chemically defined, and consequently strong foam formation could potentially be expected during cultivation. Because of the mechanical action hindering foam formation (the foam is continuously mixed into the medium by the wave action), the addition of antifoaming agents becomes unnecessary.

Noninstrumented small-scale systems, or systems with limited instrumentation, such as disposable T-flasks, spinner flasks, roller flasks, and their modifications, whose handling has been, to some extent, automated over the past few years, are regarded as routine workhorses in cell culture laboratories.

The application of noninvasive optical sensor technology to transparent cultivation containers for animal cells has resulted in highly automated or precisely monitored and/or controlled disposable microbioreactor systems. This has paved the way for a change in early-stage process development from being unmonitored to being well characterized and controlled, and has made an important contribution to the accurate replication of larger-scale conditions.

In-seed inoculum productions, process developments and GMP manufacturing processes for mAb products and vaccines, and wave-mixed and stirred-bag bioreactors are increasingly replacing fixed-wall cell culture bioreactors. Furthermore, they are displacing the early disposable bioreactors such as roller bottles, Cell Factories, and hollow fiber bioreactors. This is because the majority of animal and human cells grow serum free and in suspension, and also because cell culture bioreactor volumes are currently shrinking due to increased product titers.

When optimized cell densities and product titers must be achieved in the shortest possible time, cell culture technologists need to be willing to move away from their gold standard, that is, the use of stirring systems. In addition to highly instrumented, scalable wave-mixed and stirred single-use bioreactors, the use of shaken disposable bioreactors and novel approaches such as the PBS or the BayShake are on the increase.

It is assumed that the pharmaceutical industry's current drive toward safe, individualized medicines (e.g., personalized antibodies, functional cells for cancer, immune and tissue replacement therapies) will contribute to the continuing growth of disposable bioreactors.

Disposable bioreactors have not played an important role to date in the cultivation of cells or tissues of plant origin and microorganisms. However, plant cell biomass, secondary metabolites for pharmaceutical use, cosmetics (e.g., PhytoCELLTec products from Mibelle Biochemistry, Switzerland), and glycoproteins have already been successfully produced in satisfactory amounts in disposable bag bioreactors. They have been wave-mixed, stirred, or pneumatically agitated.

Similarly, for microorganism cultivations, where high-density growth is often desired, disposable bioreactors ensuring higher power input and oxygen transfer efficiency should be used. Currently, the user may have access to the first suitable types recommended for microorganisms, for example, the CELLtainer Microbial Bioreactor, the CellMaker Regular, or the microbial version of the XDR-Disposable Stirred Tank Bioreactor. The Nucleo Bioreactor represents another suitable bag bioreactor for microorganisms due to its high K_La values reaching 200 per hour (Figure 5.12).

The Game Changers in Disposable Bioreactor Industry

Clichés aside, every industry goes through game-changing technology, breakthrough technology, or whatever comes that wakes up the industry to new ways of doing things. In the field of bioreactors there have been four events that can be listed as "game changers."

The history of fermentation dates back to 7000–8000 BCE when the folks in Georgia and Iran began making wine (Figure 5.13). The game-changing moment did not come until a crucial experiment was carried out in 1896 by the German chemist Eduard Buchner. Buchner ground up a group of cells with sand until they were totally destroyed. He then extracted the liquid that remained and added it to a sugar solution. His assumption was that fermentation could no longer occur since the cells that had held the ferments were

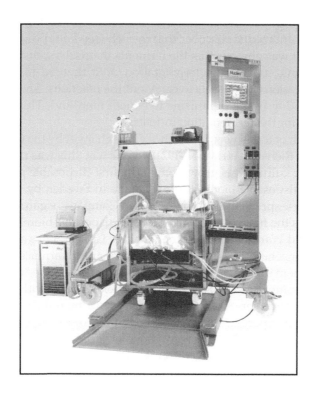

FIGURE 5.12
Nucleo bioreactor.

FIGURE 5.13
Early fermentation systems.

dead, so they no longer carried the "life force" needed to bring about fermentation. He was amazed to discover that the cell-free liquid did indeed cause fermentation. It was obvious that the ferments themselves, distinct from any living organisms, could cause fermentation. And that led to a formal process of fermentation and vessels to carry out the reactions. That changed the game of exploiting living organisms to benefit mankind. This was the first game changer for the fermentation industry.

The petri dish of today was definitely a game changer by the German bacteriologist Julius Richard Petri (1852–1921); in one way this was the first laboratory bioreactor: culture was grown on it. The first disposable petri dish (and thus the first disposable bioreactor) was made in Sweden by a little-known company; the use spread in 1960 when Sterilin Company began selling it in the 1960s. This was the first game changer for the disposable bioreactor industry.

During World War II, the governments of the United States and the United Kingdom approached the largest U.S. chemical and pharmaceutical companies to enlist them in the race to mass-produce penicillin, the "wonder drug." One of these companies, Pfizer, already had experience with fermentation techniques, first implemented 20 years earlier to manufacture citric acid. Building on that experience, Pfizer succeeded in producing large quantities of penicillin using deep-tank fermentation. Its success helped make penicillin available to Allied soldiers by the end of the war. The American Chemical Society designated the development of deep-tank fermentation by Pfizer as a National Historic Chemical Landmark on June 12, 2008, in Brooklyn, New York. This was indeed a game changer for the fermentation industry and resulted in the birth of the fermenter industry that remained true to the first experiments by Pfizer.

It was not until 1998 when Vijay Singh introduced his Wave Bioreactor using a disposable plastic bag that the industry woke up to a new method of harvesting biological engines. Even though the system was designed for cell culture and remains today useful only for cell culture, this was indeed a game changer for the disposable bioreactor industry and made the concept of disposable components acceptable to the industry.

The fourth game changer for the disposable bioreactor industry is the platform of MayaBio, in which flexible 2D plastic bags are installed on a heated flat stationary surface and a flapper pushes down at one end of the bag to create a wave motion inside the bag, which contains a proprietary ceramic sparger; this technology broke new ground:

- Use of flexible 2D bags to grow bacteria as well as every type of cell and organism in 2D bags.
- Removal of physical limitations on the size of bioreactor, since keeping the bags stationary eliminated the stress that came on the bag seams as their sizes grew.

- Daisy-chaining smaller reactors to form a larger batch that complied with CFR 21 requirements eliminated the need to qualify several batch sizes.
- Use of generic bags of all types because of the open structure of the bioreactor.
- The least expensive fully validated cGMP-compliant bioreactor in the world: cost savings on capital investment were 70%–90% compared to available systems and cost savings for ongoing components 50%+. This opened up the access of bioreactors to smaller companies, research organizations, and developing countries.

Bioreactor Brand	Vendor	Maximum Size	Main Applications
Mechanically Driven/Wave-Mixed (Horizontally Oscillation)			
BIOSTAT CultiBag RM (in the past Bio Wave)	Sartorius-Stedim	300 L CV	Cultivation of animal cells, plant cells, and microorganisms having up to medium oxygen demands: screening, seed inoculum production, small- and medium-volume-scale manufacture
Wave Bioreactor	GE Healthcare	500 L CV	Cultivation of animal cells, seed inoculum production
AppliFlex	Applikon	25 L CV	
Tsunami Bioreactor	Tsunami Bio	160 L CV per platform	No longer available
CELL-trainer Bioreactor, animal	Lonza	15 L CV	Cultivation of animal cells and plant cells: screening, seed inoculum production, sample production, small-volume-scale manufacture
CELL-trainer Bioreactor, microbial			Cultivation of microorganisms: screening, seed inoculum production, small-volume-scale manufacture
WUB	Nestlé	100 L CV	Cultivation of plant cells: small- and medium-volume-scale manufacture
Mechanically driven/ vertically Oscillation Bay Shake Bioreactor	Bayer Technology Services/ Sartorius Stedim	1,000 L TV	Cultivation of animal cells: seed inoculum production, sample production, small- and medium-volume-scale manufacture

Bioreactor Brand	Vendor	Maximum Size	Main Applications
Mechanically driven/ orbitally shaken μ24 Microbioreactor	Applikon	7 mL TV	Cultivation of animal cells, plant cells, and microorganisms: screening
BioLector	Mp2-labs	1.5 mL TV	
CulitFlask 50DBa	Sartorius-Stedim	35 mL CV	
Sensolux		1 L TV	
SB-200X Disposable Shaken Bioreactor System	Kühner/Sartorius Stedim	200 L TV	Cultivation of animal cells: seed inoculum production, sample production, small- and medium-volume-scale manufacture
CURRENT Bioreactor	AmProtein	300 L CV	
Mechanically driven/ stirred S.U.B.	Thermo Fisher Scientific	1,000 L CV	Cultivation of animal cells: seed inoculum production, small- and medium-volume-scale manufacture
BIOSTAT CultiBag STR	Sartorius Stedim	1,000 L CV	
Nucleo Bioreactor	ATMI Life Science	1,000 L CV	
XDR-DSTB, animal	Xcellerex	2,000 L CV	
XDR-DSTB, microbial		200 L TV	Manufacture of microbial HCD products
Mobius CellReady 3 L Bioreactor	Applikon/ Millipore	3 L TV	Cultivation of animal cells: screening, seed inoculum production, sample production
CelliGen BLU SUB	New Brunswick	14 L TV	
SuperSpinner D1000a	Sartorius-Stedim	1 L CV	
Pneumatically Driven			
SBB	Nestlé	100 L CV	Cultivation of plant cells: small- and medium-volume-scale manufacture
PBS	PBS	250 L TV	Cultivation of animal cells: seed inoculum production, sample production small- and medium-volume-scale manufacture
CellMaker Regular (in the past CellMaker Lite) Hybrid	Cellexus	50 L CV	Cultivation of microorganisms: seed inoculum production, sample production, small-volume-scale manufacture

Bioreactor Brand	Vendor	Maximum Size	Main Applications
CellMaker Plus	Cellexus	8 L CV	Cultivation of animal cells: seed inoculum production, sample production
MayaBio	MayaBioReactor	1 L–5000 L CV	2-D bag on a stationary platform, wave motion induced by a flapper, proprietary sparging system allows cultivation of every type of cell and organism.

Source: Eibl, R., Kaiser, S., Lombriser, R., and Eibl, D. 2010. After Appl Microbiol Biotechnol. 86: 41–49.

Appendix I is a summary of the most current publications on the design of new bioreactors.

Appendix I. Current Literature Survey of the Use of Disposable Systems

1. *Biotechnol Prog*. 2010 Oct 11. doi: 10.1002/btpr.516. Experimental characterization of flow conditions in 2- and 20-L bioreactors with wave-induced motion. Kalmbach A, Bordás R, Oncül AA, Thévenin D, Genzel Y, Reichl U. Professur für Strömungsmechanik, Bioprocess Engineering Division, Helmut-Schmidt-Universität Hamburg, D-22043 Hamburg, Germany; Max-Planck-Institut für Dynamik Komplexer Technischer Systeme, Bioprocess Engineering Division, D-39106 Magdeburg, Germany.

Quantifying the influence of flow conditions on cell viability is essential for successful control of cell growth and cell damage in major biotechnological applications, such as in recombinant protein and antibody production or vaccine manufacturing. In the last decade, new bioreactor types have been developed. In particular, bioreactors with wave-induced motion show interesting properties (e.g., disposable bags suitable for cGMP manufacturing, no requirement for cleaning and sterilization of cultivation vessels, and fast setup of new production lines) are considered in this study. As an additional advantage, it is expected that cultivation in such bioreactors result in lower shear stress compared with conventional stirred tanks. As a consequence, cell damage would be reduced as cell viability is highly sensitive to hydrodynamic conditions. To check these assumptions, an experimental setup was developed to measure the most important flow parameters (liquid surface level, liquid velocity, and liquid and wall shear stress) in two cellbag sizes (2 and 20 L) of Wave Bioreactors®. The measurements confirm in particular low shear stress values in both cellbags, indicating favorable hydrodynamic conditions for cell cultivation.

2. *Biotechnol Lett.* 2011 Jan 26. TubeSpin bioreactor 50 for the high-density cultivation of Sf-9 insect cells in suspension. Xie Q, Michel PO, Baldi L, Hacker DL, Zhang X, Wurm FM. Laboratory of Cellular Biotechnology, Institute of Bioengineering, École Polytechnique Fédérale de Lausanne, CH J2 506, Station 6, 1015, Lausanne, Switzerland.

Here we present the TubeSpin bioreactor 50 (TubeSpins) as a simple and disposable culture system for Sf-9 insect cells in suspension. Sf-9 cells had substantially better growth in TubeSpins than in spinner flasks. After inoculation with 10^6 cells/mL, maximal cell densities of 16×10^6 and 6×10^6 cells/mL were reached in TubeSpins and spinner flasks, respectively. In addition, the cell viability in these batch cultures remained above 90% for 10 days in TubeSpins but only for 4 days in spinner flasks. Inoculation at even higher cell densities reduced the duration of the lag phase. After inoculation at 2.5×10^6 cells/mL, the culture reached the maximum cell density within 3 days instead of 7 days as observed for inoculation with 10^6 cells/mL. Infection of Sf-9 cells in TubeSpins or spinner flasks with a recombinant baculovirus coding for green fluorescent protein (GFP) resulted in similar GFP-specific fluorescence levels. TubeSpins are thus an attractive option for the small-scale cultivation of Sf-9 cells in suspension and for baculovirus-mediated recombinant protein production.

3. *Chimia (Aarau).* 2010; 64(11):819–23. Innovative, nonstirred bioreactors in scales from milliliters up to 1000 L for suspension cultures of cells using disposable bags and containers—a Swiss contribution. Werner S, Eibl R, Lettenbauer C, Röll M, Eibl D, De Jesus M, Zhang X, Stettler M, Tissot S, Bürki C, Broccard G, Kühner M, Tanner R, Baldi L, Hacker D, Wurm FM. Zürcher Hochschule für Angewandte Wissenschaften, Institut für Biotechnologie, Wädenswil.

Innovative mixing principles in bioreactors, for example, using the rocking of a platform to induce a backwards and forwards "wave," or using orbital shaking to generate a wave that runs around in a cylindrical container, have proved successful for the suspension cultures of cells, especially when combined with disposable materials. This article presents an overview of the engineering characteristics when these new principles are applied in bioreactors, and case studies covering scales of operation from milliliters to 1000 L.

4. *Chimia (Aarau).* 2010; 64(11):803–7. Process monitoring with disposable chemical sensors fit in the framework of process analysis technology (PAT) for innovative pharmaceutical development and quality assurance. Spichiger S, Spichiger-Keller UE. C-CIT AG, Einsiedlerstr. 29, Wadenswil.

The innovative principle of enzymatic sensors applied to monitor the feeding process in disposable bioreactors is described. Innovative is the type of enzyme immobilized within the "paste" to monitor l-glutamate. Innovative is the application of the miniaturized disposable sensor developed at C-CIT AG for continuous monitoring. The sensor allows the amount of the digested nutrient to be estimated from the amperometric signal. Innovative is the wireless signal transduction between the sensor mounted to the bioreactor and the signal receiver. An example of a process control run is given and, also, the biocompatibility and the specifications of the biosensors. The comparison of results evaluated by different analytical methods is discussed.

5. *Biotechnol J*. 2011 Jan;6(1):56–65. A single-use purification process for the production of a monoclonal antibody produced in a PER.C6 human cell line. Kuczewski M, Schirmer E, Lain B, Zarbis-Papastoitsis G. PERCIVIA LLC, 1 Hampshire St., Cambridge, MA. mkuczewski@percivia.com

Advances in single-use technologies can enable greater speed, flexibility, and a smaller footprint for multiproduct production facilities, such as at a contract manufacturer. Recent efforts in the area of cell line and media optimization have resulted in bioreactor productivities that exceed 8 g/L in fed-batch processes or 25 g/L in high-density cell culture processes. In combination with the development of single-use stirred-tank bioreactors with larger working volumes, these intensified upstream processes can now be fit into a single-use manufacturing setting. Contrary to these upstream advances, downstream single-use technologies have been slower to follow, mostly limited by low capacity, high cost, and poor scalability. In this study, we describe a downstream process based solely on single-use technologies that meets the challenges posed by expression of a mAb (IgG(1)) in a high-density suspension culture of PER.C6 cells. The cell culture harvest was clarified by enhanced cell settling (ECS) and depth filtration. Precipitation was used for crude purification of the mAb. A high-capacity chromatographic membrane was then used in bind/elute mode, followed by two membranes in flow-through (FT) mode for polishing. A proof of concept of the entire disposable process was completed for two different scales of the purification train.

6. *Int J Artif Organs.* 2010 Aug;
33(8):512–25. Expansion of
human mesenchymal stem cells
in a fixed-bed bioreactor system
based on nonporous glass
carrier—part A: inoculation,
cultivation, and cell harvest
procedures. Weber C, Freimark
D, Pörtner R, Pino-Grace P, Pohl
S, Wallrapp C, Geigle P, Czermak
P. Institute of Biopharmaceutical
Technology, University of
Applied Sciences Giessen-
Friedberg, Giessen, Germany.

Human mesenchymal stem cells (hMSC) are a
promising cell source for several applications of
regenerative medicine. The cells employed are either
autologous or allogenic; by using stem cell lines in
particular, allogenic cells enable the production of
therapeutic cell implants or tissue-engineered
implants in stock. For these purposes, the generally
small initial cell number has to be increased; this
requires the use of bioreactors, which offer controlled
expansion of the hMSC under GMP-conform
conditions. In this study, divided into parts A and B,
a fixed-bed bioreactor system based on nonporous
borosilicate glass spheres for the expansion of hMSC,
demonstrated with the model cell line hMSC-TERT,
is introduced. The system offers convenient
automation of the inoculation, cultivation, and
harvesting procedures. Furthermore, the bioreactor
has a simple design that favors its manufacture as a
disposable unit. Part A is focused on the inoculation,
cultivation, and harvesting procedures. Cultivations
were performed in lab scales up to a bed volume of
300 cm³. The study showed that the fixed-bed system,
based on 2 mm borosilicate glass spheres, as well as
the inoculation, cultivation, and harvesting
procedures are suitable for the expansion of hMSC
with high yield and vitality.

7. *Biotechnol Prog.* 2010 Jul–Aug; 26(4):1200–3. Use of disposable reactors to generate inoculum cultures for *E. coli* production fermentations. Mahajan E, Matthews T, Hamilton R, Laird MW. Process Development Engineering, Genentech, Inc., South San Francisco, CA. ektam@gene.com

Disposable technology is being used more each year in the biotechnology industry. Disposable bioreactors allow one to avoid expenses associated with cleaning, assembly, and operations, as well as equipment validation. The Wave Bioreactor is well established for Chinese Hamster Ovary (CHO) production; however, it has not yet been thoroughly tested for *E. coli* production because of the high oxygen demand and temperature maintenance requirements of that platform. The objective of this study is to establish a robust process to generate inoculum for *E. coli* production fermentations in a Wave Bioreactor. We opted not to evaluate the WAVE system for production cultures because of the high cell densities required in our current *E. coli* production processes. Instead, the Wave Bioreactor 20/50 system was evaluated at laboratory scale (10 L) to generate inoculum with target optical densities (OD(550)) of 15 within 7–9 h (preestablished target for stainless steel fermentors). The maximum settings for rock rate (40 rpm) and angle (10.5) were used to maximize mass transfer. The gas feed was also supplemented with additional oxygen to meet the high respiratory demand of the culture. The results showed that the growth profiles for the inoculum cultures were similar to those obtained from conventional stainless steel fermentors. These inoculum cultures were subsequently inoculated into 10 L working volume stainless steel fermentors to evaluate the inocula performance of two different production systems during recombinant protein production. The results of these production cultures using WAVE inocula showed that the growth and recombinant protein production was comparable to the control data set. Furthermore, an economic analysis showed that the WAVE system would require less capital investment for installation, and operating expenses would be less than traditional stainless steel systems.

8. *Biotechnol Bioeng.* 2010 Dec 1; 107(5):802–13. Biomass production of hairy roots of *Artemisia annua* and *Arachis hypogaea* in a scaled-up mist bioreactor. Sivakumar G, Liu C, Towler MJ, Weathers PJ. Arkansas Biosciences Institute, Arkansas State University, Jonesboro, AR 72467.

Hairy roots have the potential to produce a variety of valuable small and large molecules. The mist reactor is a gas-phase bioreactor that has shown promise for low-cost culture of hairy roots. Using a newer, disposable culture bag, mist reactor performance was studied with two species, *Artemisia annua* L. and *Arachis hypogaea* (peanut), at scales from 1 to 20 L. Both species of hairy roots when grown at 1 L in the mist reactor showed growth rates that surpassed that in shake flasks. From the information gleaned at 1 L, Arachis was scaled further to 4 and then 20 L. Misting duty cycle, culture medium flow rate, and timing of when flow rate was increased were varied. In a mist reactor, increasing the misting cycle or increasing the medium flow rate are the two alternatives for increased delivery of liquid nutrients to the root bed. Longer misting cycles beyond 2–3 min were generally deemed detrimental to growth. On the other hand, increasing the medium flow rate to the sonic nozzle, especially during the exponential phase of root growth (weeks 2–3), was the most important factor for increasing growth rates and biomass yields in the 20 L reactors. *A. hypogaea* growth in 1 L reactors was $\mu = 0.173$ day(-1) with a biomass yield of 12.75 g DW L(-1). This exceeded that in shake flasks at $\mu = 0.166$ day(-1) and 11.10 g DW L(-1). Best growth rate and biomass yield at 20 L was $\mu = 0.147$ and 7.77 g DW L(-1), which was mainly achieved when medium flow rate delivery was increased. The mist deposition model was further evaluated using this newer reactor design, and when the apparent thickness of roots (+hairs) was taken into account, the empirical data correlated with model predictions. Together these results establish the most important conditions to explore for future optimization of the mist bioreactor for the culture of hairy roots.

9. *Biotechnol Bioeng*. 2010 Aug 15; 106(6):906–17. Production of cell culture (MDCK) derived live attenuated influenza vaccine (LAIV) in a fully disposable platform process. George M, Farooq M, Dang T, Cortes B, Liu J, Maranga L. Cell Culture Development, MedImmune, 3055 Patrick Henry Dr., Santa Clara, California 95054. georgem@medimmune.com

The majority of influenza vaccines are manufactured using embryonated hens' eggs. The potential occurrence of a pandemic outbreak of avian influenza might reduce or even eliminate the supply of eggs, leaving the human population at risk. Also, the egg-based production technology is intrinsically cumbersome and not easily scalable to provide a rapid worldwide supply of vaccine. In this communication, the production of a cell culture (Madin-Darby canine kidney (MDCK)) derived live attenuated influenza vaccine (LAIV) in a fully disposable platform process using a novel Single Use Bioreactor (SUB) is presented. The cell culture and virus infection was maintained in a disposable stirred tank reactor with PID control of pH, DO, agitation, and temperature, similar to traditional glass or stainless steel bioreactors. The application of this technology was tested using MDCK cells grown on microcarriers in proprietary serum-free medium and infection with 2006/2007 seasonal LAIV strains at 25–30 L scale. The MDCK cell growth was optimal at the agitation rate of 100 rpm. Optimization of this parameter allowed the cells to grow at a rate similar to that achieved in the conventional 3 L glass stirred-tank bioreactors. Influenza vaccine virus strains, A/New Caledonia/20/99 (H1N1 strain), A/Wisconsin/67/05 (H3N2 strain), and B/Malaysia/2506/04 (B strain) were all successfully produced in SUB with peak virus titers > or =8.6 log(10) FFU/mL. This result demonstrated that more than 1 million doses of vaccine can be produced through one single run of a small bioreactor at the scale of 30 L and thus provided an alternative to the current vaccine production platform with fast turnaround and low up-front facility investment, features that are particularly useful for emerging and developing countries and clinical trial material production.

10. *Biotechnol Prog*. 2010 Sep;
26(5):1431–7. Rapid protein
production using CHO stable
transfection pools. Ye J, Alvin K,
Latif H, Hsu A, Parikh V,
Whitmer T, Tellers M, de la Cruz
Edmonds MC, Ly J, Salmon P,
Markusen JF. Merck & Co., Inc.,
Bioprocess Research and
Development, Rahway, NJ 07065.
jianxin_ye@merck.com

During early preclinical development of therapeutic proteins, representative materials are often required for process development, such as for pharmacokinetic/pharmacodynamic studies in animals, formulation design, and analytical assay development. To rapidly generate large amounts of representative materials, transient transfection is commonly used. Because of the typical low yields with transient transfection, especially in CHO cells, here we describe an alternative strategy using stable transfection pool technology. Using stable transfection pools, gram quantities of monoclonal antibody (mAb) can be generated within 2 months posttransfection. Expression levels for monoclonal antibodies can be achieved ranging from 100 mg/L to over 1000 mg/L. This methodology was successfully scaled up to a 200 L scale using disposable bioreactor technology for ease of rapid implementation. When fluorescence-activated cell sorting was implemented to enrich the transfection pools for high producers, the productivity could be improved by about threefold. We also found that an optimal production time window exists to achieve the highest yield because the transfection pools were not stable and productivity generally decreased over length in culture. The introduction of Universal chromatin-opening elements into the expression vectors led to significant productivity improvement. The glycan distribution of the mAb product generated from the stable transfection pools was comparable to that from the clonal stable cell lines.

11. *Biotechnol Bioeng*. 2010 Oct 15; 107(3):497–505. Microfluidic biolector-microfluidic bioprocess control in microtiter plates. Funke M, Buchenauer A, Schnakenberg U, Mokwa W, Diederichs S, Mertens A, Müller C, Kensy F, Büchs J. AVT Biochemical Engineering, RWTH Aachen University, Worringerweg 1, 52074 Aachen, Germany.

In industrial-scale biotechnological processes, the active control of the pH value combined with the controlled feeding of substrate solutions (fed-batch) is the standard strategy to cultivate both prokaryotic and eukaryotic cells. On the contrary, for small-scale cultivation, much simpler batch experiments with no process control are performed. This lack of process control often hinders researchers in scaling up and scaling down fermentation experiments, because the microbial metabolism, and thereby the growth and production kinetics, drastically changes depending on the cultivation strategy applied. While small-scale batches are typically performed highly parallel and in high-throughput, large-scale cultivations demand sophisticated equipment for process control, which is in most cases costly and difficult to handle. Currently, there is no technical system on the market that realizes simple process control in high throughput. The novel concept of a microfermentation system described in this work combines a fiber-optic online monitoring device for microtiter plates (MTPs)—the BioLector technology—together with microfluidic control of cultivation processes in volumes below 1 mL. In the microfluidic chip, a micropump is integrated to realize distinct substrate flow rates during fed-batch cultivation in microscale. Hence, a cultivation system with several distinct advantages could be established: (1) high information output on a microscale; (2) many experiments can be performed in parallel and be automated using MTPs; (3) this system is user-friendly and can easily be transferred to a disposable single-use system. This article elucidates this new concept and illustrates applications in fermentations of *Escherichia coli* under pH-controlled and fed-batch conditions in shaken MTPs.

12. *Biotech Histochem*. 2010 Aug; 85(4):213–29. Tissue engineered tumor models. Ingram M, Techy GB, Ward BR, Imam SA, Atkinson R, Ho H, Taylor CR. Huntington Medical Research Institutes, 99 North El Molino Avenue, Pasadena, CA 91101-1830.

Many research programs use well-characterized tumor cell lines as tumor models for in vitro studies. Because tumor cells grown as three-dimensional (3D) structures have been shown to behave more like tumors in vivo than do cells growing in monolayer culture, a growing number of investigators now use tumor cell spheroids as models. Single-cell-type spheroids, however, do not model the stromal–epithelial interactions that have an important role in controlling tumor growth and development in vivo. We describe here a method for generating, reproducibly, more realistic 3D tumor models that contain both stromal and malignant epithelial cells with an architecture that closely resembles that of tumor microlesions in vivo. Because they are so tissue-like, we refer to them as tumor histoids. They can be generated reproducibly in substantial quantities. The bioreactor developed to generate histoid constructs is described and illustrated. It accommodates disposable culture chambers that have filled volumes of either 10 or 64 mL, each culture yielding on the order of 100 or 600 histoid particles, respectively. Each particle is a few tenths of a millimeter in diameter. Examples of histological sections of tumor histoids representing cancers of breast, prostate, colon, pancreas, and urinary bladder are presented. Potential applications of tumor histoids include, but are not limited to, use as surrogate tumors for prescreening antisolid tumor pharmaceutical agents, as reference specimens for immunostaining in the surgical pathology laboratory, and use in studies of invasive properties of cells or other aspects of tumor development and progression. Histoids containing nonmalignant cells also may have potential as "seeds" in tissue engineering. For drug testing, histoids probably will have to meet certain criteria of size and tumor cell content. Using a COPAS Plus flow cytometer, histoids containing fluorescent tumor cells were analyzed successfully and sorted using such criteria.

13. *Bioprocess Biosyst Eng.* 2010 Oct; 33(8):961–70. Epub 2010 Mar 27. A BOD monitoring disposable reactor with alginate-entrapped bacteria. Villalobos P, Acevedo CA, Albornoz F, Sánchez E, Valdés E, Galindo R, Young ME. Centro de Biotecnología, Universidad Técnica Federico Santa María, Avenida España 1680, Valparaíso, Chile. patricio. villalobos@usm.cl

Biochemical oxygen demand (BOD) is a measure of the amount of dissolved oxygen that is required for the biochemical oxidation of the organic compounds in 5 days. New biosensor-based methods have been conducted for a faster determination of BOD. In this study, a mathematical model to evaluate the feasibility of using a BOD sensor, based on disposable alginate-entrapped bacteria, for monitoring BOD in situ was applied. The model considers the influences of alginate bead size and bacterial concentration. The disposable biosensor can be adapted according to specific requirements depending on the organic load contained in the wastewater. Using Klein and Washausen parameter in a Lineweaver–Burk plot, the glucose diffusivity was calculated in $6.4 \times 10(-10)$ (m^2/s) for beads 1 mm in diameter, and slight diffusion restrictions were observed ($n = 0.85$). Experimental results showed a correlation ($p < 0.05$) between the respirometric peak and the standard BOD test. The biosensor response was representative of BOD.

14. *Protein Pept Lett.* 2010 Jul;17(7):919–24. Transient expression of recombinant sPDGFR alpha-Fc in CHO DG44 cells using 50-mL orbitally shaking disposable bioreactors. Sang YX, Zhang XW, Chen XJ, Xie K, Qian CW, Hong A, Xie QL, Xiong S. Biomedical R&D Center, Guangdong Provincial Key Laboratory of Bioengineering Medicine, National Engineering Research Center of Genetic Medicine, Jinan University, Guangzhou 510632, Guangdong, PR China.

Overactivity of platelet-derived growth factor (PDGF) has been linked to malignant cancers. High levels of PDGF result in the activation of its receptors (PDGFRs) and the overproliferation of cells. Therefore, interfering with this signaling pathway in cancer cells could be significant for anticancer drug development. In a previous study, the sPDGFR alpha-Fc fusion protein expressed in static CHO-k(1) cells showed an antiproliferative effect on vascular endothelial cells. However, it was difficult to obtain a large quantity of this fusion protein for further functional studies. In the present study, the sPDGFR alpha-Fc fusion protein was transiently expressed in Chinese Hamster Ovary (CHO) DG44 cells in 50 mL orbital shaking bioreactors. sPDGFR alpha-Fc was expressed as a 250 kDa dimeric protein with potential glycosylation. The final yield of sPDGFR alpha-Fc in the culture supernatant was as high as 16.68 mg/L. Our results suggest that transient expression in orbital shaking bioreactors may be feasible for preparation of recombinant proteins used for preclinical studies.

15. *Appl Microbiol Biotechnol.* 2010
Mar;86(1):41–9. Epub 2010 Jan 22.
Disposable bioreactors: the
current state-of-the-art and
recommended applications in
biotechnology. Eibl R, Kaiser S,
Lombriser R, Eibl D. School of
Life Sciences and Facility
Management, Institute of
Biotechnology, Zurich University
of Applied Sciences, P.O. Box,
CH-8820, Wädenswil,
Switzerland. regine.eibl@zhaw.ch

Disposable bioreactors have increasingly been incorporated into preclinical, clinical, and production-scale biotechnological facilities over the last few years. Driven by market needs, and, in particular, by the developers and manufacturers of drugs, vaccines, and further biologicals, there has been a trend toward the use of disposable seed bioreactors as well as production bioreactors. Numerous studies documenting their advantages in use have contributed to further new developments and have resulted in the availability of a multitude of disposable bioreactor types that differ in power input, design, instrumentation, and scale of the cultivation container. In this review, the term *disposable bioreactor* is defined, the benefits and constraints of disposable bioreactors are discussed, and critical phases and milestones in the development of disposable bioreactors are summarized. An overview of the disposable bioreactors that are currently commercially available is provided, and the domination of wave-mixed, orbitally shaken, and, in particular, stirred disposable bioreactors in animal-cell-derived productions at cubic meter scale is reported. The growth of this type of reactor system is attributed to the recent availability of stirred disposable benchtop systems such as the Mobius CellReady 3 L Bioreactor. Analysis of the data from computational fluid dynamic simulation studies and first cultivation runs confirms that this novel bioreactor system is a viable alternative to traditional cell culture bioreactors at benchtop scale.

16. Bioresour Technol. 2010
Apr;101(8):2896–9. Epub 2009
Dec 31. Study on
poly-hydroxyalkanoate (PHA)
production in pilot scale
continuous mode wastewater
treatment system. Chakravarty P,
Mhaisalkar V, Chakrabarti T. Ion
Exchange Waterleau Ltd.,
Process and Proposal Division,
Reveira Apartments, 4th Floor,
Plot No. 134, 6-3-347/9
Punjagutta, Hyderabad 500 082,
India. partha_chakravarty@
hotmail.com

Generation of poly-hydroxyalkanoates (PHAs) from milk and ice-cream processing wastewater was studied in a continuous mode reactor system at pilot scale. The integrated system comprised an anaerobic acidogenic reactor (AAR), a conventional activated sludge production reactor (ASPR), and a PHA synthesis reactor (PHAR) to induce PHA accumulation in the biomass, which was finally harvested while treating the raw dairy wastewater to meet the disposal limits, thereby reducing generation of disposable sludge. The PHA content in the PHA rich biomass was approximately 43% of the sludge dry weight. Kinetics of both ASPR and PHAR were studied. The maximum PHA yield coefficient ($Y(sp)$ (max)) with respect to COD degradation in the PHAR was derived as 0.25 kg PHA/kg of COD degraded. Similarly, the kinetic parameters, that is, $K(s)$, micro(m), $Y(obs)$, and $k(d)$ of the ASPR were 37.16 mg/L COD, 0.97 d(-1), 0.51 mg MLSS/mg COD, and 0.049 d(-1), respectively.

17. *Biotechnol Prog.* 2010 Mar; 26(2):332–51. Technological progresses in monoclonal antibody production systems. Rodrigues ME, Costa AR, Henriques M, Azeredo J, Oliveira R. IBB-Institute for Biotechnology and Bioengineering, Centre of Biological Engineering, University of Minho, Campus de Gualtar 4710-057 Braga, Portugal.

Monoclonal antibodies (mAbs) have become vitally important to modern medicine and are currently one of the major biopharmaceutical products in development. However, the high clinical dose requirements of mAbs demand a greater biomanufacturing capacity, leading to the development of new technologies for their large-scale production, with mammalian cell culture dominating the scenario. Although some companies have tried to meet these demands by creating bioreactors of increased capacity, the optimization of cell culture productivity in normal bioreactors appears to be a better strategy. This review describes the main technological progress made with this idea, presenting the advantages and limitations of each production system, as well as suggestions for improvements. New and upgraded bioreactors have emerged both for adherent and suspension cell culture, with disposable reactors attracting increased interest in the last few years. Furthermore, the strategies and technologies used to control culture parameters are in constant evolution, aiming at the online multiparameter monitoring and considering how parameters were not seen as relevant for process optimization in the past. All progress being made has as primary goal the development of highly productive and economic mAb manufacturing processes that will allow the rapid introduction of the product in the biopharmaceutical market at more accessible prices.

18. *Appl Microbiol Biotechnol.* 2010 Feb;85(5):1339–51. Epub 2009 Dec 3. Bench to batch: advances in plant cell culture for producing useful products. Weathers PJ, Towler MJ, Xu J. Department of Biology and Biotechnology at Gateway, Worcester Polytechnic Institute, Worcester, MA 01609. weathers@wpi.edu

Despite significant efforts over nearly 30 years, only a few products produced by in vitro plant cultures have been commercialized. Some new advances in culture methods and metabolic biochemistry have improved the useful potential of plant cell cultures. This review will provide references to recent relevant reviews along with a critical analysis of the latest improvements in plant cell culture, co-cultures, and disposable reactors for production of small secondary product molecules, transgenic proteins, and other products. Some case studies for specific products or production systems are used to illustrate principles.

19. *Microb Cell Fact.* 2009 Aug 5;8:44. Comparisons of optically monitored small-scale stirred tank vessels to optically controlled disposable bag bioreactors. Hanson MA, Brorson KA, Moreira AR, Rao G. Center for Advanced Sensor Technology, Chemical and Biochemical Engineering Department, University of Maryland Baltimore County, Baltimore, MD, 21250. grao@umbc.edu.

Upstream bioprocesses are extremely complex since living organisms are used to generate active pharmaceutical ingredients (APIs). Cells in culture behave uniquely in response to their environment; thus, culture conditions must be precisely defined and controlled in order for productivity and product quality to be reproducible. Thus, development culturing platforms are needed where many experiments can be carried out at once and pertinent scale-up information can be obtained. Here we have tested a high-throughput bioreactor (HTBR) as a scale-down model for a lab-scale wave-type bioreactor (CultiBag). Mass transfer was characterized in both systems and scaling based on volumetric oxygen mass transfer coefficient (K_La) was sufficient to give similar DO trends. HTBR and CultiBag cell growth and mAb production were highly comparable in the first experiment, where DO and pH were allowed to vary freely. In the second experiment, growth and mAb production rates were lower in the HTBR as compared to the CultiBag, where pH was controlled. The differences in magnitude were not considered significant for biological systems. Similar oxygen delivery rates were achieved in both systems, leading to comparable culture performance (growth and mAb production) across scales and mode of mixing. The HTBR model was most fitting when neither system was pH-controlled, providing an information-rich alternative to typically nonmonitored mL-scale platforms.

20. *Adv Biochem Eng Biotechnol.* 2010; 115:1–31. Disposable bioreactors: maturation into pharmaceutical glycoprotein manufacturing. Brecht R. ProBioGen AG, Goethestrasse 54, Berlin, Germany. rene.brecht@ probiogen.de

Modern biopharmaceutical development is characterized by deep understanding of the structure–activity relationship of biological drugs. Therefore, the production process has to be tailored more to the product requirements than to the existing equipment in a certain facility. In addition, the major challenges for the industry are to lower the high production costs of biologics and to shorten the overall development time. The flexibility for providing different modes of operation using disposable bioreactors in the same facility can fulfill these demands and support tailor-made processes. Over the last 10 years, a huge and still increasing number of disposable bioreactors have entered the market. Bioreactor volumes of up to 2,000 L can be handled by using disposable bag systems. Each individual technology has been made available for different purposes up to the GMP compliant production of therapeutic drugs, even for market supply. This chapter summarizes disposable technology development over the last decade by comparing the different technologies and showing trends and concepts for the future.

21. *Adv Biochem Eng Biotechnol.* 2010; 115:33–53. Use of orbital shaken disposable bioreactors for mammalian cell cultures from the milliliter-scale to the 1,000-liter scale. Zhang X, Stettler M, De Sanctis D, Perrone M, Parolini N, Discacciati M, De Jesus M, Hacker D, Quarteroni A, Wurm F. Laboratory of Cellular Biotechnology, Ecole Polytechnique Fédérale de Lausanne, CH-1015, Lausanne, Switzerland.

Driven by the commercial success of recombinant biopharmaceuticals, there is an increasing demand for novel mammalian cell culture bioreactor systems for the rapid production of biologicals that require mammalian protein processing. Recently, orbitally shaken bioreactors at scales from 50 mL to 1,000 L have been explored for the cultivation of mammalian cells and are considered to be attractive alternatives to conventional stirred-tank bioreactors because of increased flexibility and reduced costs. Adequate oxygen transfer capacity was maintained during the scale-up, and strategies to increase further oxygen transfer rates (OTR) were explored, while maintaining favorable mixing parameters and low-stress conditions for sensitive lipid-membrane-enclosed cells. Investigations from process development to the engineering properties of shaken bioreactors are under way, but the feasibility of establishing a robust, standardized, and transferable technical platform for mammalian cell culture based on orbital shaking and disposable materials has been established with further optimizations and studies ongoing.

22. *Adv Biochem Eng Biotechnol.* 2010; 115:117–43. Transport advances in disposable bioreactors for liver tissue engineering. Catapano G, Patzer JF 2nd, Gerlach JC. Department of Chemical Engineering and Materials, University of Calabria, Rende (CS), Italy, catapano@unical.it

Acute liver failure (ALF) is a devastating diagnosis with an overall survival of approximately 60%. Liver transplantation is the therapy of choice for ALF patients but is limited by the scarce availability of donor organs. The prognosis of ALF patients may improve if essential liver functions are restored during liver failure by means of auxiliary methods, because liver tissue has the capability to regenerate and heal. Bioartificial liver (BAL) approaches use liver tissue or cells to provide ALF patients with liver-specific metabolism and synthesis products necessary to relieve some of the symptoms and to promote liver tissue regeneration. The most promising BAL treatments are based on the culture of tissue-engineered (TE) liver constructs, with mature liver cells or cells that may differentiate into hepatocytes to perform liver-specific functions, in disposable continuous-flow bioreactors. In fact, adult hepatocytes perform all essential liver functions. Clinical evaluations of the proposed BALs show that they are safe, but their treatment efficacy is not clearly proved as compared to standard supportive treatments. Ambiguous clinical results, the time loss of cellular activity during treatment, and the presence of a necrotic core in the cell compartment of many bioreactors suggest that improvement of transport of nutrients, and metabolic wastes and products to or from the cells in the bioreactor is critical for the development of therapeutically effective BALs. In this chapter, advanced strategies that have been proposed over to improve mass transport in the bioreactors at the core of a BAL for the treatment of ALF patients are reviewed.

23. *Appl Microbiol Biotechnol.* 2009 Jul; 83(5):809–23. Epub 2009 Jun 2. Bioprocessing of plant cell cultures for mass production of targeted compounds. Georgiev MI, Weber J, Maciuk A. Department of Microbial Biosynthesis and Biotechnologies, Institute of Microbiology, Bulgarian Academy of Sciences, Plovdiv, Bulgaria. milengeorgiev@gbg.bg

More than a century has passed since the first attempt to cultivate plant cells in vitro. During this time, plant cell cultures have become an increasingly attractive and cost-effective alternatives to classical approaches for the mass production of plant-derived metabolites. Furthermore, plant cell culture is the only economically feasible way of producing some high-value metabolites (e.g., paclitaxel) from rare and threatened plants. This review summarizes recent advances in bioprocessing aspects of plant cell cultures, from callus culture to product formation, with particular emphasis on the development of suitable bioreactor configurations (e.g., disposable reactors) for plant-cell-culture-based processes; the optimization of bioreactor culture environments as a powerful means of improving yields; bioreactor operational modes (fed-batch, continuous, and perfusion); and biomonitoring approaches. Recent trends in downstream processing are also considered.

24. *Adv Biochem Eng Biotechnol*. 2010; 115:89–115. Disposable bioreactors for plant micropropagation and mass plant cell culture. Ducos JP, Terrier B, Courtois D. Nestlé R&D Centre Tours, 101 Avenue Gustave Eiffel, Notre Dame D'Oé, BP 49716, 37097, Tours Cedex 2, France.

Different types of bioreactors are used at Nestlé R&D Centre - Tours for mass propagation of selected plant varieties by somatic embryogenesis and for large-scale culture of plants cells to produce metabolites or recombinant proteins. Recent studies have been directed to cut down the production costs of these two processes by developing disposable cell culture systems. Vegetative propagation of elite plant varieties is achieved through somatic embryogenesis in liquid medium. A pilot-scale process has recently been set up for the industrial propagation of *Coffea canephora* (Robusta coffee). The current production capacity is 3.0 million embryos per year. The pregermination of the embryos was previously conducted by temporary immersion in liquid medium in 10 L glass bioreactors. An improved process has been developed using a 10 L disposable bioreactor consisting of a bag containing a rigid plastic box ("Box-in-Bag" bioreactor), ensuring, among other advantages, a higher light transmittance to the biomass due to its horizontal design. For large-scale cell culture, two novel flexible plastic-based disposable bioreactors have been developed from 10 to 100 L working volumes, validated with several plant species ("Wave and Undertow" and "Slug Bubble" bioreactors). The advantages and the limits of these new types of bioreactor are discussed, based mainly on our own experience on coffee somatic embryogenesis and mass cell culture of soya and tobacco.

25. *Adv Biochem Eng Biotechnol*. 2010; 115:145–69. Sensors in disposable bioreactors status and trends. Glindkamp A, Riechers D, Rehbock C, Hitzmann B, Scheper T, Reardon KF. Institute for Technical Chemistry, Leibniz University Hannover, Callinstr. 3, 30167, Hannover, Germany, glindkamp@iftc.uni-hannover.de.

For better control of productivity and product quality, detailed monitoring of various parameters is required. Since disposable bioreactors are becoming more and more important for biotechnological applications, adequate sensors for this type of reactor are necessary. The required properties of sensors used in disposable reactors differ from those of sensors for multiuse reactors. For example, sensors that are in direct contact with the medium must be inexpensive, but do not need a long lifetime, since they can be used only once. This chapter gives an overview of the state of the art and future trends in the field of sensors suited for use in disposable bioreactors. The main focus here is on in situ sensors, which can be based on optical, semiconductor, and ultrasonic technologies, but current concepts for disposable sampling units are also reviewed.

26. *Hum Gene Ther.* 2009
Aug;20(8):861–70. Scalable
recombinant adeno-associated
virus production using
recombinant herpes simplex
virus type 1 coinfection of
suspension-adapted mammalian
cells. Thomas DL, Wang L,
Niamke J, Liu J, Kang W, Scotti
MM, Ye GJ, Veres G, Knop DR.
Applied Genetic Technologies
Corporation, Alachua, FL 32615.

Recombinant adeno-associated virus (rAAV) production systems capable of meeting clinical or anticipated commercial-scale manufacturing needs have received relatively little scrutiny compared with the intense research activity afforded the in vivo and in vitro evaluation of rAAV for gene transfer. Previously, we have reported a highly efficient recombinant herpes simplex virus type 1 (rHSV) complementation system for rAAV production in multiple adherent cell lines; however, production in a scalable format was not demonstrated. Here, we report rAAV production by rHSV coinfection of baby hamster kidney (BHK) cells grown in suspension (sBHK cells), using two ICP27-deficient rHSV vectors, one harboring a transgene flanked by the AAV2 inverted terminal repeats and a second bearing the AAV rep2 and capX genes (where X is any rAAV serotype). The rHSV coinfection of sBHK cells produced similar rAAV1/AAT-specific yields (85,400 DNase-resistant particles [DRP]/cell) compared with coinfection of adherent HEK-293 cells (74,600 DRP/cell); however, sBHK cells permitted a threefold reduction in the rHSV-rep2/capX vector multiplicity of infection, grew faster than HEK-293 cells, retained specific yields (DRP/cell) at higher cell densities, and had a decreased virus production cycle. Furthermore, sBHK cells were able to produce AAV serotypes 1, 2, 5, and 8 at similar specific yields, using multiple therapeutic genes. rAAV1/AAT production in sBHK cells was scaled to 10 L disposable bioreactors, using optimized spinner flask infection conditions, and resulted in average volumetric productivities as high as $2.4 \times 10(14)$ DRP/liter.

27. *Adv Biochem Eng Biotechnol.* 2009 Apr 17. Application of disposable bag-bioreactors in tissue engineering and for the production of therapeutic agents. Eibl R, Eibl D. Zurich University of Applied Sciences, School of Life Sciences and Facility Management, Institute of Biotechnology, Campus Grüntal, Wädenswil, , CH-8820, Switzerland. regine.eibl@zhaw.ch

In order to increase process efficiency, many pharmaceutical and biotechnology companies have introduced disposable bag technology over the last 10 years. Because this technology also greatly reduces the risk of cross-contamination, disposable bags are preferred in applications in which an absolute or improved process safety is a necessity, namely the production of functional tissue for implantation (tissue engineering), the production of human cells for the treatment of cancer and immune system diseases (cellular therapy), the production of viruses for gene therapies, the production of therapeutic proteins, and veterinary as well as human vaccines. Bioreactors with a presterile cultivation bag made of plastic material are currently used in both development and manufacturing processes primarily operating with animal and human cells at small- and middle-volume scale. Due to their scalability, hydrodynamic expertise and the convincing results of oxygen transport efficiency studies, wave-mixed bioreactors are the most used, together with stirred-bag bioreactors and static bags, which have the longest tradition. Starting with a general overview of disposable bag bioreactors and their main applications, the following paper summarizes the working principles and engineering aspects of bag bioreactors suitable for cell expansion, formation of functional tissue and production of therapeutic agents. Furthermore, results from selected cultivation studies are presented and discussed.

28. *Gene Ther.* 2009 Jun; 16(6):766–75. Epub 2009 Apr 2. Purification of recombinant baculoviruses for gene therapy using membrane processes. Vicente T, Peixoto C, Carrondo MJ, Alves PM. IBET/ ITQB-UNL, Oeiras, Portugal.

Recombinant baculoviruses (rBVs) are widely used as vectors for the production of recombinant proteins in insect cells. More recently, these viral vectors have been gaining increasing attention due to their emerging potential as gene therapy vehicles to mammalian cells. Their production in stirred bioreactors using insect cells is an established technology; however, the downstream processing (DSP) of baculoviruses envisaged for clinical applications is still poorly developed. In the present work, the recovery and purification of rBVs aiming at injectable-grade virus batches for gene therapy trials was studied. A complete downstream process comprising three steps—depth filtration, ultra/ diafiltration, and membrane sorption—was successfully developed. Optimal operational conditions for each individual step were achieved, yielding a scalable DSP for rBVs as vectors for gene therapy. The processing route designed hereby presents global recovery yields reaching 40% (at purities over 98%) and, most importantly, relies on technologies easy to transfer to process scales under cGMP guidelines.

29. *Cytotechnology.* 1999 Jul; 30(1–3):149–58. Disposable bioreactor for cell culture using wave-induced agitation. Singh V. Schering-Plough Research Institute, 1011 Morris Avenue, Union, NJ, 07083.

This work describes a novel bioreactor system for the cultivation of animal, insect, and plant cells using wave agitation induced by a rocking motion. This agitation system provides good nutrient distribution, off-bottom suspension, and excellent oxygen transfer without damaging fluid shear or gas bubbles. Unlike other cell culture systems, such as spinners, hollow-fiber bioreactors, and roller bottles, scale-up is simple, and has been demonstrated up to 100 L of culture volume. The bioreactor is disposable, and therefore requires no cleaning or sterilization. Additions and sampling are possible without the need for a laminar flow cabinet. The unit can be placed in an incubator requiring minimal instrumentation. These features dramatically lower the purchase cost, and operating expenses of this laboratory/pilot scale cell cultivation system. Results are presented for various model systems: (1) recombinant NS0 cells in suspension; (2) adenovirus production using human 293 cells in suspension; (3) Sf9 insect cell/baculovirus system; and (4) human 293 cells on microcarrier. These examples show the general suitability of the system for cells in suspension, anchorage-dependent culture, and virus production in research and GMP applications.

30. *Biotechnol J*. 2008 Oct; 3(9–10):1185–200. Implementation of advanced technologies in commercial monoclonal antibody production. Zhou JX, Tressel T, Yang X, Seewoester T. Process and Product Development, Amgen, Inc., Thousand Oaks, CA 91320-1799. joez@amgen.com

Process advancements driven through innovations have been key factors that enabled successful commercialization of several human therapeutic antibodies in recent years. The production costs of these molecules are higher in comparison to traditional medicines. In order to lower the development and later manufacturing costs, recent advances in antibody production technologies target higher-throughput processes with increased clinical and commercial economics. In this review, essential considerations and trends for commercial process development and optimization are described, followed by the challenges to obtain a high titer cell culture process and its subsequent impact on the purification process. One of these recent technical advances is the development and implementation of a disposable Q membrane adsorber as an alternative to a Q-packed-bed column in a flow-through mode. The scientific concept and principles underlining Q membrane technology and its application are also reviewed.

31. *J Biotechnol*. 2008 Jun 30; 135(3):272–80. Epub 2008 May 21. Expression of SEAP (secreted alkaline phosphatase) by baculovirus mediated transduction of HEK 293 cells in a hollow fiber bioreactor system. Jardin BA, Zhao Y, Selvaraj M, Montes J, Tran R, Prakash S, Elias CB. Biotechnology Research Institute, 6100 Royalmount Avenue, Montreal, Quebec, Canada.

A BacMam baculovirus was designed in our laboratory to express the reporter protein secreted alkaline phosphatase (SEAP) driven by the immediate early promoter of human cytomegalovirus promoter (CMV). In vitro tests have been carried out using this recombinant baculovirus to study the secreted protein in two cell lines and under various culture conditions. The transductions were carried out on two commonly used mammalian cell lines, namely, the human embryonic kidney (HEK 293A) and Chinese hamster ovary (CHO-K1). Initial studies clearly demonstrated that the transient expression of SEAP was at least tenfold higher in the HEK 293 cells than the CHO cells under equivalent experimental conditions. Factorial design experiments were done to study the effect of different parameters such as cell density, multiplicity of infection (MOI), and the histone deacetylase inhibitor, trichostatin A concentration. The MOI and the cell density were found to have the most impact on the process. The enhancer trichostatin A also showed some positive effect. The production of secreted protein in a batch reactor was studied using the Wave disposable bioreactor system. A semicontinuous perfusion process was developed to extend the period of gene expression in mammalian cells using a hollow fiber bioreactor system (HFBR). The growth of cells and viability in both systems was monitored by offline analyses of metabolites. The expression of recombinant protein could be maintained over an extended period of time up to 30 days in the HFBR.

32. *Tissue Eng.* 2007 Dec;
13(12):3003–10. Bioreactor for
application of subatmospheric
pressure to three-dimensional
cell culture. Wilkes RP, McNulty
AK, Feeley TD, Schmidt MA,
Kieswetter K. Kinetic Concepts
Inc., San Antonio, Texas 78249.
wilkesr@lci1.com

Vacuum-assisted closure (VAC) negative pressure
wound therapy (NPWT) is a highly successful and
widely used treatment modality for wound healing,
although no apparatus exists to monitor the effects of
subatmospheric pressure application in vitro. Such
an apparatus is desirable to better understand the
biological effects of this therapy and potentially
improve upon them. This article describes the
development and validation of a novel bioreactor
that permits such study. Tissue analogues consisting
of 3-dimensional fibroblast-containing fibrin clots
were cultured in off-the-shelf disposable cell culture
inserts and multiwell plates that were integrated into
the bioreactor module. Negative pressure dressings,
commercialized for wound therapy, were placed on
top of the culture, and subatmospheric pressure was
applied to the dressing. Cultures were perfused with
media at controlled physiologic wound exudate flow
rates. The design of this bioreactor permits
observation of the culture using an inverted
microscope in brightfield and fluorescence modes
and sustained incubation of the system in a 5%
carbon dioxide atmosphere. This closed-system
mimics the wound microenvironment under VAC
NPWT. Matrix compression occurs as the
subatmospheric pressure draws the dressing material
down. At the contact zone, surface undulations were
clearly evident on the fibroblast-containing tissue
analogues at 24 h and appeared to correspond to the
dressing microstructure. The bioreactor design,
consisting of sterilizable machined plastics and
disposable labware, can be easily scaled to multiple
units. Validation experiments show that cell survival
in this system is comparable with that seen in cells
grown in static tissue culture. After application of
VAC NPWT, cell morphology changed, with cells
appearing thicker and with an organized actin
cytoskeleton. The development and validation of this
new culture system establishes a stable platform for
in vitro investigations of subatmospheric pressure
application to tissues.

33. *Conf Proc IEEE Eng Med Biol Soc.* 2006; 1:632–5. Multiple automated minibioreactor system for multifunctional screening in biotechnology. Fontova A, Soley A, Gálvez J, Sarró E, Lecina M, Rosell J, Riu P, Cairó J, Gòdia F, Bragós R. Electronic Engineering Dept., Technical University of Catalonia, Barcelona, Spain.

The current techniques applied in biotechnology allow to obtain many types of molecules that must be tested on cell cultures (high-throughput screening HTS). Although such tests are usually carried out automatically on mini- or microwell plates, the procedures in the preindustrial stage are performed almost manually on higher-volume recipients known as bioreactors. The growth conditions in both stages are completely different. The screening system presented in this work is based on the multiwell test plates philosophy, a disposable multiple minibioreactor that allows reproduction of industrial bioreactor culture conditions: aeration, stirring, temperature, O_2, pH, and visible-range optical absorbance measurements. It is possible to reproduce the growth conditions for both suspended and adherent animal cell types using 1 to 10 mL vol. bioreactors. In the case of bacteria or yeast, it is not possible to achieve a high biomass concentration, due to the reduced head volume air supply.

34. *Biotechnol Adv.* 2008 Jan–Feb;
26(1):46–72. Epub 2007 Sep 19.
Upstream processes in antibody
production: evaluation of critical
parameters. Jain E, Kumar A.
Department of Biological
Sciences and Bioengineering,
Indian Institute of Technology
Kanpur, 208016-Kanpur, India.

The demand for monoclonal antibodies for therapeutic and diagnostic applications is rising constantly, which brings about a need to bring down the cost of its production. In this context, it becomes a prerequisite to improving the efficiency of the existing processes used for monoclonal antibody production. This review describes various upstream processes used for monoclonal antibody production and evaluates critical parameters and efforts that are being made to enhance the efficiency of the process. The upstream technology has been greatly upgraded from host cells used for manufacturing to bioreactors type and capacity. The host cells used a range from microbial, mammalian, to plant cells, with mammalian cells dominating the scenario. Disposable bioreactors are being promoted for small-scale production due to easy adaptation to process validation and flexibility, though they are limited by the scale of production. In this respect, Wave Bioreactors for suspension culture have been introduced recently. A novel bioreactor for immobilized cells is described that permits an economical and easy alternative to the hollow fiber bioreactor at lab-scale production. Modification of the cellular machinery to alter its metabolic characteristics has further added to robustness of cells and perks up cell specific productivity. The process parameters, including feeding strategies and environmental parameters, are being improved, and efforts to validate them to get reproducible results are becoming a trend. Online monitoring of the process and product characterization is increasingly gaining importance. In total, the advancement of upstream processes have led to an increase in volumetric productivity by 100-fold over the last decade and make monoclonal antibody production a more economical and realistic option for therapeutic applications.

35. *Biotechnol Prog*. 2007 Nov–Dec; 23(6):1340–6. Epub 2007 Oct 3. Novel orbital shake bioreactors for transient production of CHO derived IgGs. Stettler M, Zhang X, Hacker DL, De Jesus M, Wurm FM. Laboratory of Cellular Biotechnology, Faculty of Life Sciences, Ecole Polytechnique Fédérale de Lausanne, CH-1015 Lausanne, Switzerland.

Large-scale transient gene expression in mammalian cells is being developed for the rapid production of recombinant proteins for biochemical and preclinical studies. Here, the scalability of transient production of a recombinant human antibody in Chinese hamster ovary (CHO) cells was demonstrated in orbitally shaken disposable bioreactors at scales from 50 mL to 50 L. First, a small-scale multiparameter approach was developed to optimize the poly(ethylenimine)-mediated transfection in 50 mL shake tubes. This study confirmed the benefit, both in terms of extended cell culture viability and increased product yield, of mild hypothermic cultivation conditions for transient gene expression in CHO cells. Second, the scalability of the process was demonstrated in disposable shake bioreactors having nominal volumes of 5, 20, and 50 L with final antibody yields between 30 and 60 mg L(-1). Thus, the combination of transient gene expression with disposable shake bioreactors allows for rapid and cost-effective production of recombinant proteins in CHO cells.

36. *Biotechnol Lett*. 2008 Feb; 30(2):253–8. Epub 2007 Sep 22. Bioreactor for solid-state cultivation of Phlebiopsis gigantea. Virtanen V, Nyyssölä A, Vuolanto A, Leisola M, Seiskari P. Laboratory of Bioprocess Engineering, Helsinki University of Technology, Espoo, Finland. veera.virtanen@kcl.fi

Phlebiopsis gigantea fungus used in biological control of root rot is currently cultivated commercially in disposable, sterilizable plastic bags. A novel-packed bed bioreactor was designed for cultivating *P. gigantea* and compared to the plastic bag method and to a tray bioreactor. The spore viability of 5.4×10^6 c.f.u./g obtained with the packed bed bioreactor was of the same order of magnitude as the viabilities obtained with the other cultivation methods. Furthermore, the packed bed bioreactor was less time- and space-consuming and easier to operate than the tray bioreactor.

37. *Biotechnol Annu Rev.* 2007;
13:95–113. Advances in antibody
manufacturing using
mammalian cells. Morrow KJ Jr.
Newport Biotechnology
Consultants, 625 Washington
Avenue, Newport, KY 41071.
kjohnmorrowjr@insightbb.com

In this review, we describe recent advances in
antibody processing technology, including: (1)
development of proprietary cell lines; (2) improved
expression systems optimized by selective
technologies to boost underperformers; (3) improved
protein-free and serum-free culture media; and (4)
attention to glycosylation and other posttranslational
modifications. Advances in computer technology
and sophisticated redesign of bioreactors have been
major contributors to the dramatic improvements in
antibody yields that have been documented in the
last decade. Disposable bioreactor components are
now widespread, resulting in improved yields,
better-quality product, and lower costs for
producers. Downstream innovations include (1)
disposable devices for clarification and purification,
(2) improved resins and ligands, and (3) new designs
of hardware for improved performance. While there
are numerous factors contributing to the increased
yields that have been obtained, the most sustained of
these is the introduction of disposable technologies
on both the upstream and the downstream ends of
the process. With the continuing introduction of
improved computer technology and technological
innovation, there is every reason to believe that the
quality and quantity of antibody products will
continue to improve in the coming years, and supply
will be adequate to meet the forthcoming needs of
the industry.

38. *J Virol Methods.* 2007 Nov; 145(2):155–61. Epub 2007 Jul 2. Production of recombinant adeno-associated vectors using two bioreactor configurations at different scales. Negrete A, Kotin RM. Laboratory of Biochemical Genetics, National Heart, Lung, and Blood Institute, US National Institutes of Health, Bethesda, MD 20892.

The conventional methods for producing recombinant adeno-associated virus (rAAV) rely on transient transfection of adherent mammalian cells. To gain acceptance and achieve current good manufacturing process (cGMP) compliance, a clinical-grade rAAV production process should have the following qualities: simplicity, consistency, cost-effectiveness, and scalability. Currently, the only viable method for producing rAAV in large scale, for example, > or =10(16) particles per production run, utilizes baculovirus expression vectors (BEVs) and insect cells suspension cultures. The previously described rAAV production in 40 L culture using a stirred-tank bioreactor requires special conditions for implementation and operation that are not available in all laboratories. Alternatives to producing rAAV in stirred-tank bioreactors are single-use, disposable bioreactors, for example, Wave. The disposable bags are purchased presterilized, thereby eliminating the need for end-user sterilization and also avoiding cleaning steps between production runs, thus facilitating the production process. In this study, rAAV production in stirred tank and Wave Bioreactors was compared. The working volumes were 10 and 40 L for the stirred-tank bioreactors and 5 and 20 L for the Wave Bioreactors. Comparable yields of rAAV, approximately 2E+13 particles per liter of cell culture, were obtained in all volumes and configurations. These results demonstrate that producing rAAV in large scale using BEVs is reproducible, scalable, and independent of the bioreactor configuration.

39. *Bioprocess Biosyst Eng.* 2007 Jul; 30(4):231–41. Evaluation of a novel Wave Bioreactor cellbag for aerobic yeast cultivation. Mikola M, Seto J, Amanullah A. Fermentation and Cell Culture, Merck and Co., P.O. Box 4, Mailstop WP26C-1, West Point, PA 19486. mark_mikola@merck. com

The Wave Bioreactor is widely used in cell culture due to the benefits of disposable technology and ease of use. A novel cellbag was developed featuring a frit sparger to increase the system's oxygen transfer. The purpose of this work was to evaluate the sparged cellbag for yeast cultivation. Oxygen mass transfer studies were conducted in simulated culture medium, and the sparged system's maximum oxygen mass transfer coefficient (K_La) was 38 h(-1). These measurements revealed that the sparger was ineffective in increasing the oxygen transfer capacity. Cultures of *Saccharomyces cerevisiae* were successfully grown in oxygen-blended sparged and oxygen-blended standard cellbags. Under steady-state conditions for both cellbag designs, K_La values as high as 60 h(-1) were obtained with no difference in growth characteristics. This is the first report of a successful cultivation of a microbe in a Wave Bioreactor, compared to conventional seed expansion in shake flasks and stirred-tank bioreactors.

40. *J Biosci Bioeng.* 2007 Jan; 103(1):50–9. Cholesterol delivery to NS0 cells: challenges and solutions in disposable linear low-density polyethylene-based bioreactors. Okonkowski J, Balasubramanian U, Seamans C, Fries S, Zhang J, Salmon P, Robinson D, Chartrain M. Merck Research Laboratories, Bioprocess R&D, PO Box 2000, RY80Y-105, Rahway, NJ 07065.

We report the successful cultivation of cholesterol-dependent NS0 cells in linear low-density polyethylene (LLDPE) Wave Bioreactors when employing a low ratio of cyclodextrin to the cholesterol additive mixture. While cultivation of NS0 cells in Wave Bioreactors was successful when using a culture medium supplemented with fetal bovine serum (FBS), cultivation with the same culture medium supplemented with cholesterol–lipid concentrate (CLC), which contains lipids and synthetic cholesterol coupled with the carrier methyl-beta-cyclodextrin (mbetaCD), proved to be problematic. However, it was possible to cultivate NS0 cells in the medium supplemented with CLC when using conventional cultivation vessels such as disposable polycarbonate shake flasks and glass bioreactors. A series of experiments investigating the effect of the physical conditions in Wave Bioreactors (e.g., rocking rate/angle, gas delivery mode) ruled out their likely influence, while the exposure of the cells to small squares of Wave Bioreactor film resulted in a lack of growth as in the Wave Bioreactor, suggesting an interaction between the cells, the CLC, and the LLDPE contact surface. Further experiments with both cholesterol-independent and cholesterol-dependent NS0 cells established that the concurrent presence of mbetaCD in the culture medium and the LLDPE film was sufficient to inhibit growth for both cell types. By reducing the excess mbetaCD added to the culture medium, it was possible to successfully cultivate cholesterol-dependent NS0 cells in Wave Bioreactors using a cholesterol–mbetaCD complex as the sole source of exogenous cholesterol. We propose that the mechanism of growth inhibition involves the extraction of cholesterol from cell membranes by the excess mbetaCD in the medium, followed by the irreversible adsorption or entrapment of the cholesterol–mbetaCD complexes to the LLDPE surface of the Wave Bioreactor. Controlling and mitigating these negative interactions enabled the routine utilization of disposable bioreactors for the cultivation of cholesterol-dependent NS0 cell lines in conjunction with an animal-component-free cultivation medium.

41. *Biotechnol Prog.* 2007 Jan-Feb; 23(1):46–51. Cell culture process development: advances in process engineering. Heath C, Kiss R. Amgen, Seattle, Washington 98119, and Genentech, South San Francisco, California 94080. kiss.robert@gene.com

Representatives from the cell culture process development community met on September 11 and 12, 2006 at the ACS National Meeting in San Francisco to discuss "Cell Culture Process Development: Advances in Process Engineering." This oral session was held as part of the Division of Biochemical Technology (BIOT) program. The presentations addressed the very small scale (less than 1 mL) to the very large scale (20,000 L). The topics covered included development of high-throughput cell culture screening systems, modeling and characterization of bioreactor environments from mixing and shear perspectives at both small and large scales, systematic approaches for improving scale-up and scale-down activities, development of disposable bioreactor technologies, and novel perfusion culture approaches. All told, this well-attended session resulted in a valuable exchange of technical information and demonstrated a high level of interest within the process development community.

42. *Open Biomed Eng J.* 2007 Oct 29; 1:64–70. Cultivation and differentiation of encapsulated hMSC-TERT in a disposable small-scale syringe-like fixed bed reactor. Weber C, Pohl S, Pörtner R, Wallrapp C, Kassem M, Geigle P, Czermak P. Institute of Biopharmaceutical Technology, University of Applied Sciences Giessen-Friedberg, Giessen-Germany.

The use of commercially available plastic syringes is introduced as disposable small-scale fixed-bed bioreactors for the cultivation of implantable therapeutic cell systems on the basis of an alginate-encapsulated human mesenchymal stem cell line. The system introduced is fitted with a noninvasive oxygen sensor for the continuous monitoring of the cultivation process. Fixed-bed bioreactors offer advantages in comparison to other systems due to their ease of automation and online monitoring capability during the cultivation process. These benefits combined with the advantage of single use make the fixed-bed reactor an interesting option for GMP processes. The cultivation of the encapsulated cells in the fixed-bed bioreactor system offered vitalities and adipogenic differentiation similar to well-mixed suspension cultures.

43. *J Struct Funct Genomics*. 2006 Jun;7(2):101–8. Epub 2006 Dec 23. Economical parallel protein expression screening and scale-up in *Escherichia coli*. Brodsky O, Cronin CN. Department of Structural Biology, Pfizer Global Research and Development, 10628 Science Center Drive, La Jolla, CA 92121.

A novel microfermentation and scale-up platform for parallel protein production in *Escherichia coli* is described. The vertical shaker device Vertiga, which generates low-volume high density (A(600) approximately 20) *Escherichia coli* cultures in 96-position deep-well plates without auxiliary oxygen supplementation, has been coupled to a new disposable shake flask design, the Ultra Yield flask, that allows for equally high cell culture densities to be obtained. The Ultra Yield flask, which accommodates up to 1 l in culture volume, has a baffled base and a more vertical wall construction compared to traditional shake flask designs. Experimental data is presented demonstrating that the Ultra Yield flask generates, on average, an equivalent amount of recombinant protein per unit cell culture density as do traditional shake flask designs but at a substantially greater amount per unit volume. The combination of Vertiga and the Ultra Yield flask provides a convenient and scalable low-cost solution to parallel protein production in *Escherichia coli*.

44. *Appl Microbiol Biotechnol*. 2007 Feb; 74(2):324–30. Epub 2006 Nov 30. Study of the oxygen transfer in a disposable flexible bioreactor with surface aeration in vibrated medium. Kilani J, Lebeault JM. Laboratoire Génie des Procédés Industriels UMR CNRS 6067, Département Génie Chimique, Université de Technologie de Compiègne, Centre de Recherche de Royallieu, B. P 20529, 60205, Compiègne Cedex, France. jacem.kilani@utc.fr

The oxygen mass transfer is a critical design parameter for most bioreactors. It can be described and analyzed by means of the volumetric mass transfer coefficient $K_L a$. This coefficient is affected by many factors such as geometrical and operational characteristics of the vessels, type, media composition, rheology, and microorganism's morphology and concentration. In this study, we aim to develop and characterize a new culture system based on the surface aeration of a flexible, single-used bioreactor fixed on a vibrating table. In this context, the K(L)a was evaluated using a large domain of operating variables such as vibration frequency of the table, overpressure inside the pouch, and viscosity of the liquid. A novel method for K(L)a determination based on the equilibrium state between oxygen uptake rate and oxygen transfer rate of the system at given conditions was also developed using resting cells of baker's fresh yeast with a measured oxygen uptake rate of 21 mg g(-1) h(-1) (at 30°C). The effect of the vibration frequency on the oxygen transfer performance was studied for frequencies ranging from 15 to 30 Hz, and a maximal K(L)a of 80 h(-1) was recorded at 30 Hz. A rheological study of the medium added with carboxymethylcellulose at different concentrations and the effect of the liquid viscosity on K(L)a were determined. Finally, the mixing time of the system was also measured using the pH method.

45. *Biosens Bioelectron.* 2007 Apr 15; 22(9–10):2071–8. Epub 2006 Oct 13. Design and development of a highly stable hydrogen peroxide biosensor on screen printed carbon electrode based on horseradish peroxidase bound with gold nanoparticles in the matrix of chitosan. Tangkuaram T, Ponchio C, Kangkasomboon T, Katikawong P, Veerasai W. Department of Chemistry, Faculty of Science, Mahidol University, Bangkok 10400, Thailand. pengtangkua@yahoo. com

The design and development of a screen-printed carbon electrode (SPCE) on a polyvinyl chloride substrate as a disposable sensor is described. Six configurations were designed on silk screen frames. The SPCEs were printed with four inks: silver ink as the conducting track, carbon ink as the working and counter electrodes, silver/silver chloride ink as the reference electrode, and insulating ink as the insulator layer. Selection of the best configuration was done by comparing slopes from the calibration plots generated by the cyclic voltammograms at 10, 20, and 30 mM $K_3Fe(CN)_6$ for each configuration. The electrodes with similar configurations gave similar slopes. The 5th configuration was the best electrode that gave the highest slope. Modifying the best SPCE configuration for use as a biosensor, horseradish peroxidase (HRP) was selected as a biomaterial bound with gold nanoparticles (AuNP) in the matrix of chitosan (HRP/AuNP/CHIT). Biosensors of HRP/SPCE, HRP/CHIT/SPCE and HRP/AuNP/CHIT/SPCE were used in the amperometric detection of H_2O_2 in a solution of 0.1M citrate buffer, pH 6.5, by applying a potential of -0.4 V at the working electrode. All the biosensors showed an immediate response to H_2O_2. The effect of HRP/AuNP incorporated with CHIT (HRP/AuNP/CHIT/SPCE) yielded the highest performance. The amperometric response of HRP/AuNP/CHIT/SPCE retained over 95% of the initial current of the 1st day up to 30 days of storage at 4°C. The biosensor showed a linear range of 0.01–11.3 mM H_2O_2, with a detection limit of 0.65 microM H_2O_2 (S/N=3). The low detection limit, long storage life, and wide linear range of this biosensor make it advantageous in many applications, including bioreactors and biosensors.

46. *Biotechnol Bioeng*. 2007 Apr 1; 96(5):914–23. Two new disposable bioreactors for plant cell culture: the wave and undertow bioreactor and the slug bubble bioreactor. Terrier B, Courtois D, Hénault N, Cuvier A, Bastin M, Aknin A, Dubreuil J, Pétiard V. Centre de Recherche & Développement Nestlé-Tours, 101, Avenue Gustave Eiffel, BP 49716, 37097 Tours Cedex 2, France. benedicte.terrier@rdto.nestle.com

The present article describes two novel flexible plastic-based disposable bioreactors. The first one, the wave and undertow (WU) bioreactor, is based on the principle of a wave and undertow mechanism that provides agitation while offering convenient mixing and aeration to the plant cell culture contained within the bioreactor. The second is a high-aspect-ratio bubble column bioreactor, where agitation and aeration are achieved through the intermittent generation of large-diameter bubbles, "Taylor-like" or "slug bubbles" (SB bioreactor). It allows an easy volume increase from a few liters to larger volumes up to several hundred liters with the use of multiple units. The cultivation of tobacco and soya cells producing isoflavones is described up to 70 and 100 L working volume for the SB Bubble Column (SB) bioreactor and WU bioreactor, respectively. The bioreactors are disposable and presterilized before use, so cleaning, sterilization, and maintenance operations are greatly reduced or eliminated. Both bioreactors represent efficient and low-cost cell culture systems, applicable to various cell cultures at small and medium scale, complementary to traditional stainless steel bioreactors.

47. J Biosci. 2006 Sep;31(3):363–8. Aujeszky's disease virus production in disposable bioreactor. Slivac I, Srcek VG, Radosevic K, Kmetic I, Kniewald Z. Laboratory for Cell Culture Technology and Biotransformations, University of Zagreb, 6 Pierotti St., HR-10000 Zagreb, Croatia.

A novel, disposable-bag bioreactor system that uses wave action for mixing and transferring oxygen was evaluated for BHK 21 C13 cell line growth and Aujeszky's disease virus (ADV) production. Growth kinetics of BHK 21 C13 cells in the Wave Bioreactor during the 3-day period were determined. At the end of the 3-day culture period and cell density of $1.82 \times 10(6)$ cells mL^{-1}, the reactor was inoculated with 9 mL of gE- Bartha K-61 strain ADV suspension $(10(5.9)$ TCID50) with multiplicity of infection (MOI) of 0.01. After a 144 h incubation period, 400 mL of ADV harvest was obtained with a titer of $10(7.0)$ TCID 50 mL^{-1}, which corresponds to 40,000 doses of vaccine against AD. In conclusion, the results obtained with the Wave Bioreactor using BHK 21 C13 cells showed that this system can be considered suitable for ADV or BHK 21 C13 cell biomass production.

48. *Biotechnol Bioeng.* 2006 Oct 5; 95(2):226–61. Bioprocess monitoring and computer control: key roots of the current PAT initiative. Junker BH, Wang HY. Bioprocess Research and Development, Merck Research Laboratories, Rahway, New Jersey 07065. beth_junker@merck.com

This review article was written for the journal, *Biotechnology and Bioengineering*, to commemorate the 70th birthday of Daniel I.C. Wang, who served as doctoral thesis advisor to each of the co-authors, but a decade apart. Key roots of the current PAT initiative in bioprocess monitoring and control are described, focusing on the impact of Danny Wang's research as a professor at MIT. The history of computer control and monitoring in biochemical processing has been used to identify the areas that have already benefited and those that are most likely to benefit in the future from PAT applications. Past applications have included the use of indirect estimation methods for cell density, expansion of online/at-line and online/in situ measurement techniques, and development of models and expert systems for control and optimization. Future applications are likely to encompass additional novel measurement technologies, measurements for multiscale and disposable bioreactors, real-time batch release, and more efficient data utilization to achieve process validation and continuous improvement goals. Dan Wang's substantial contributions in this arena have been one key factor in steering the PAT initiative toward realistic and attainable industrial applications.

49. *Lab Chip.* 2006 Sep; 6(9):1229–35.
Epub 2006 Jul 27.
Microbioreactor arrays with
integrated mixers and fluid
injectors for high-throughput
experimentation with pH and
dissolved oxygen control. Lee
HL, Boccazzi P, Ram RJ, Sinskey
AJ. Research Laboratory of
Electronics, Massachusetts
Institute of Technology, 77
Massachusetts Ave, Cambridge,
MA 02139.

We have developed an integrated array of microbioreactors, with 100 microL working volume, comprising a peristaltic oxygenating mixer and microfluidic injectors. These integrated devices were fabricated in a single chip and can provide a high oxygen transfer rate (k(L)a approximately 0.1 s(-1)) without introducing bubbles, and closed-loop control over dissolved oxygen and pH (±0.1). The system was capable of supporting eight simultaneous *Escherichia coli* fermentations to cell densities greater than 13 g-dcw L(-1) (1 cm OD(650 nm) > 40). This cell density was comparable to that achieved in a 4 L reference fermentation, conducted with the same strain, in a bench-scale stirred-tank bioreactor and is more than four times higher than cell densities previously achieved in microbioreactors. Bubble-free oxygenation permitted near-real-time optical density measurements that could be used to observe subtle changes in the growth rate and infer changes in the state of microbial genetic networks. Our system provides a platform for the study of the interaction of microbial populations with different environmental conditions, which has applications in basic science and industrial bioprocess development. We leverage the advantages of microfluidic integration to deliver a disposable, parallel bioreactor in a single chip, rather than robotically multiplexing independent bioreactors, which opens a new avenue for scaling small-scale bioreactor arrays with the capabilities of bench-scale stirred-tank reactors.

50. *Protein Expr Purif.* 2006 Dec;50(2):185–95. Epub 2006 Jul 4. A semi-automated large-scale process for the production of recombinant tagged proteins in the Baculovirus expression system. Schlaeppi JM, Henke M, Mahnke M, Hartmann S, Schmitz R, Pouliquen Y, Kerins B, Weber E, Kolbinger F, Kocher HP. Discovery Technologies, Biomolecules Production Unit, Novartis Institutes for BioMedical Research, Bdg. WSJ-508.2.21, CH-4002 Basel, Switzerland. jean-marc. schlaeppi@novartis.com

The efficient preparation of recombinant proteins at the lab-scale level is essential for drug discovery, in particular for structural biology, protein interaction studies, and drug screening. The Baculovirus insect-cell expression system is one of the most widely applied and highly successful systems for production of recombinant functional proteins. However, the use of eukaryotic cells as host organisms and the multistep protocol required for the generation of sufficient virus and protein has limited its adaptation to industrialized high-throughput operation. We have developed an integrated large-scale process for continuous and partially automated protein production in the Baculovirus system. The instrumental platform includes parallel insect-cell fermentation in 10L BioWave reactors, cell harvesting and lysis by tangential flow filtration (TFF) using two custom-made filtration units, and automated purification by multidimensional chromatography. The use of disposable materials (bags, filters, and tubing), automated cleaning cycles, and column regeneration prevent any cross-contamination between runs. The preparation of the clear cell lysate by sequential TFF takes less than 2 h and represents considerable time saving compared to standard cell harvesting and lysis by sonication and ultracentrifugation. The process has been validated with 41 His-tagged proteins with molecular weights ranging from 20 to 160 kDa. These proteins represented several families, and included 23 members of the deubiquitinating enzyme (DUB) family. Each downstream unit can process four proteins in less than 24 h with final yields between 1 and 100 mg, and purities between 50% and 95%.

51. *Biochem Biophys Res Commun*.
2006 Jun 30; 345(2):602–7. Epub
2006 Apr 19. Expression of a
human anti-rabies virus
monoclonal antibody in tobacco
cell culture. Girard LS, Fabis MJ,
Bastin M, Courtois D, Pétiard V,
Koprowski H. Thomas Jefferson
University, 1020 Locust Street,
Biotechnology Foundation,
Philadelphia, PA 19107.

A *Nicotiana tabacum* cv. Xanthi cell culture was
initiated from a transgenic plant expressing a human
antirabies virus monoclonal antibody. Within 3
months, plant cell suspension cultures were
established, and recombinant protein expression was
examined. The antibody was stably produced during
culture growth. ELISA, protein G purification,
Western blotting, and neutralization assay confirmed
that the antibody was fully processed, with
association of light and heavy chains, and that it was
able to bind and neutralize rabies virus.
Quantification of antibody production in plant cell
suspension culture revealed 30 microg/g of cell dry
weight for the highest-producing culture (0.5 mg/L),
3 times higher than from the original transgenic
plant. The same production level was observed 3
months after cell culture initiation. Plant cell
suspension cultures were successfully grown in a
new disposable plastic bioreactor, with a growth rate
and production level similar to that of cultures in
Erlenmeyer flasks.

52. *Biophys J*. 2006 May 15;
90(10):3813–22. Epub 2006 Feb
24. Soft trapping and
manipulation of cells using a
disposable nanoliter biochamber.
Diop M, Taylor R. Department of
Physics, The University of
Western Ontario, London,
Ontario, Canada.

Low-power continuous-wave laser radiation is used
to form a very stable microbubble at the end of a
specially etched and metalized optical fiber probe.
We demonstrate that the microbubble, which is
firmly attached to the fiber probe, can be used to
benignly trap and manipulate living swine sperm
cells as well as human embryonic kidney cells. The
lifetime of the microbubble has been prolonged and
the gaseous environment inside the bubble
controlled using micropipette gas injection. The
controlled fusion of two microbubbles is
demonstrated as a means of transferring
microparticles from one bubble to another. These
experiments lay the foundation for the use of the
microbubble as a mobile, nanoliter-volume
disposable biochamber for cellular studies.

53. *Biosens Bioelectron.* 2005 Sep 15; 21(3):445–54. Epub 2004 Dec 21. Semi disposable reactor biosensors for detecting carbamate pesticides in water. Suwansa-ard S, Kanatharana P, Asawatreratanakul P, Limsakul C, Wongkittisuksa B, Thavarungkul P. Biophysics Research Unit: Biosensors and Biocurrents, Prince of Songkla University, Hat Yai, Songkhla 90112, Thailand.

Two flow-injection biosensor systems using a semidisposable enzyme reactor have been developed to determine carbamate pesticides in water samples. Acetylcholinesterase was immobilized on silica gel by covalent binding. pH and conductivity electrodes were used to detect the ionic change of the sample solution due to hydrolysis of acetylcholine. Carbamate pesticides inhibited acetylcholinesterase, and the decrease in the enzyme activity was used to determine these pesticides. Parameters influencing the performance of the systems were optimized to be used in the inhibition procedure. Carbofuran and carbaryl were used to test these systems. Detection limits for the potentiometric and conductimetric systems were both at 10% inhibition corresponding to 0.02 and 0.3 ppm of carbofuran and carbaryl, respectively. Both systems also provided the same linear ranges, 0.02–8.0 ppm for carbofuran, and 0.3–10 ppm for carbaryl. The analysis of pesticides was done a few times before the reactor was disposed. Percentages of inhibition obtained from different reactors were reproducible; therefore, no recalibration was necessary when changing the reactor. The biosensors were used to analyze carbaryl in water samples from six wells in a vegetable-growing area. Both systems could detect the presence of carbaryl in the samples and provided good recoveries of the added carbaryl, that is, 80%–106% for the potentiometric system and 75%–105% for the conductimetric system. The presence of carbaryl in water samples analyzed by the biosensors was confirmed by the gas chromatography–mass spectrometric system. These biosensors do not require any sample preconcentration and are suitable for detecting pesticides in real water samples.

54. *Protein Expr Purif.* 2003 Jun; 29(2):311–20. A less laborious approach to the high-throughput production of recombinant proteins in *Escherichia coli* using 2-liter plastic bottles. Millard CS, Stols L, Quartey P, Kim Y, Dementieva I, Donnelly MI. Environmental Research Division, Argonne National Laboratory, Bldg. 202/Rm. BE111, 9700 South Cass Avenue, Argonne, IL 60439.

Contemporary approaches to biology often call for the high-throughput production of large amounts of numerous proteins for structural or functional studies. Even with the highly efficient protein expression systems developed in *Escherichia coli*, production of these proteins is laborious and time consuming. We have simplified established protocols by the use of disposable culture vessels: common 2 L polyethylene terephthalate beverage bottles. The bottles are inexpensive, fit conveniently in commonly available flask holders, and, because they are notched, provide sufficient aeration to support the growth of high-density cultures. The use of antibiotics and freshly prepared media alleviates the need for sterilization of media and significantly reduces the labor involved. Uninoculated controls exhibited no growth during the time required for protein expression in experimental cultures. The yield, solubility, activity, and pattern of crystallization of proteins expressed in bottles were comparable to those obtained under conventional culture conditions. After use, the bottles are discarded, reducing the risk of cross-contamination of subsequent cultures. The approach appears to be suitable for high-throughput production of proteins for structural or functional studies.

55. *Biotechnol Prog.* 2003 Jan–Feb; 19(1):2–8. Fluid mechanics, cell distribution, and environment in CellCube bioreactors. Aunins JG, Bader B, Caola A, Griffiths J, Katz M, Licari P, Ram K, Ranucci CS, Zhou W. Merck Research Laboratories, West Point, Pennsylvania.

Cultivation of MRC-5 cells and attenuated hepatitis A virus (HAV) for the production of VAQTA, an inactivated HAV vaccine (1), is performed in the CellCube reactor, a laminar flow fixed-bed bioreactor with an unusual diamond-shaped, diverging-converging flow geometry. These disposable bioreactors have found some popularity for the production of cells and gene therapy vectors at intermediate scales of operation (2, 3). Early testing of the CellCube revealed that the fluid mechanical environment played a significant role in nonuniform cell distribution patterns generated during the cell growth phase. Specifically, the reactor geometry and manufacturing artifacts, in combination with certain inoculum practices and circulation flow rates, can create cell growth behavior that is not simply explained. Through experimentation and computational fluid dynamics simulations, we can account for practically all of the observed cell growth behavior, which appears to be due to a complex mixture of flow distribution, particle deposition under gravity, fluid shear and, possibly, nutritional microenvironment.

56. *Int J Artif Organs.* 1999 Dec; 22(12):816–22. Rotary cell culture system (RCCS): a new method for cultivating hepatocytes on microcarriers. Mitteregger R, Vogt G, Rossmanith E, Falkenhagen D. Christian Doppler Laboratory for Specific Adsorption Technologies in Medicine, Krems, Austria. mitteregger@zbmt.donau-uni.ac.at

The Rotary Cell Culture System (RCCS) is a new technology for growing anchorage-dependent or suspension cells in the laboratory. The RCCS is a horizontally rotated, bubble-free disposable culture vessel with diffusion gas exchange. The system provides a reproducible, complex 3D in vitro culture system with large cell masses. During cell growing, the rotation speed can be adjusted to compensate for increased sedimentation rates. The unique environment of low shear forces, high mass transfer, and microgravity provides very good cultivating conditions for many cell types, cell aggregates, or tissue particles in a standard tissue culture laboratory. The system enables culturing of HepG2 cells on Cytodex 3 microcarriers (mcs) to high densities. We inoculated $2 \times 10(5)$/mL HepG2 cells and 200 mg Cytodex 3 mcs in 50 mL Williams E medium (incl. 10% fetal calf serum (FCS), allowing them to attach to the mcs in the rotating vessel (rotation rate 14–20 rpm). HepG2 cells readily attached to the mcs while the vessel was rotating. Attachment of HepG2 to the mcs was about 50% after 24 h and 100% within 48 h. After 72 h of rotary culturing, small aggregates of Hep G2 on mcs were built. HepG2 cells and the aggregates rotated with the vessel and did not settle within the vessel or collide with the wall of the vessel. We conclude that this new RCCS is an excellent technology for culturing HepG2 cells on Cytodex 3 mcs. The system is easy to handle and enables culturing of anchorage-dependent cells to high densities in a short period.

6

Connectors and Transfers

Angels are spirits, flames of fire; they are higher than man, they have wider connections.

Matthew Simpson

Disposable components came into use first in the field of connectors and lines, as it was difficult to clean them. Unlike hard piping, the flexible tubing incorporated into disposable transfer lines does not require costly and time-consuming cleaning and validation. This allows manufacturers to quickly change process steps or convert over to a new product. This is a key advantage for multiple product facilities in which process requirements change depending on the drug being produced. Innovative manufacturers now incorporate disposable tubing assemblies throughout the bioprocess from seed trains to final fill applications. Additional cost savings result from reduced labor, chemical, water, and energy demands associated with cleaning and validation.

Modern bioprocessing facilities scale up inoculum from a few million cells in several milliliters of culture to production volumes of thousands of liters. This process requires aseptic transfer at each point along the seed train. Traditional bioprocessing facilities accomplish scale-up using a dedicated series of stainless steel bioreactors linked together with valves and rigid tubing. For these systems, to prevent contamination between production runs, a clean-in-place (CIP) system is designed into each bioreactor, vessel, and piping line to remove any residual materials. These CIP and steam-in-place (SIP) systems require extensive validation testing, and the valves and piping contained in these systems can create additional validation challenges.

Advances in disposable technology allow bioprocess engineers to replace most storage vessels and fixed-piping networks with disposable storage systems and tubing assemblies. Disposables eliminate the need for CIP validation for many components and reduce maintenance and capital expense by eliminating expensive vessels, valves, and sanitary piping assemblies.

While total disposable systems are not always possible, there is a transition taking place and often there is a need to connect a disposable system with stainless steel vessels. Disposable media storage systems are routinely manufactured for volumes from 20 to 2,500 L. Media storage systems arrive at the bioprocess facility sterilized by gamma irradiation and often are fitted with integrated filters, sampling systems, and connectors. Using an SIP connector such as Colder Products' Steam-Thru® Connection (www.colder.com) allows operators to make sterile connections between these presterilized disposable systems and stainless steel bioreactors for aseptic transfer of media.

Similarly, disposable tubing assemblies may be used to transfer inoculum between bioreactors using either a peristaltic pump or headspace pressure. Such transfer lines can reduce the number of reusable valves required for transfer and eliminate problem areas for CIP and SIP validation. Terminating each presterilized transfer line with a disposable SIP connector provides sterility assurance equal to that of traditional fixed piping at lower capital costs.

As disposable bioreactors are beginning to appear, companies are using them for both seed trains and small-scale production. These systems are connected to a cell culture media storage bag (either by aseptic welding or aseptic connectors such as Colder Products' AseptiQuik®) using flexible tubing. Flexible tubing with aseptic connectors is used as transfer lines between each reactor in the process.

There are also instances when liquids are transferred from a higher-ISO environment to a lower-ISO environment, and assurance is needed that it does not result in cross-contamination; to ensure this, a conduit can be installed in the walls connecting the two areas, with the cleaner room having a higher pressure. A presterilized tube is then inserted from the lower ISO class side to the higher ISO class side and connected to the vessels between which the liquid is transferred by a peristaltic pump; upon completion of transfer, the tube is pulled into the higher-ISO class area and discarded. This method allows connection between downstream and upstream areas without the risk of transferring any contamination to a lower-ISO class area such as a downstream area.

Tubing

Flexible tubes are an essential part of all disposable systems and are subject to the safety concerns described in an earlier chapter with regard to the leachables and extractables. Several attributes of flexible tubing require evaluation such as their heat resistance, operating temperature range, chemical resistance, color, density, shore hardness, flexibility, elasticity, surface smoothness, mechanical stability, abrasion resistance, gas permeability, visible and ultraviolet (UV) light sensitivity, composition of layers, weldability, sealability, and sterilizability by gamma radiation or in an autoclave.

All tubes used in bioprocessing conform to USP Class VI classification, FDA 21 CFR 177.2600, and EP 3A Sanitary Standard. For cGMP manufacturing, these are classified as bulk pharmaceuticals. The most common materials are used for the tubing include

- Thermoplastic elastomer (C-Flex, PharmaPure, PharMed BPT, SaniPur 60, Advanta Flex) is pump tubing, highly biocompatible with easy seal-ability and low permeability. Thermoplastic tubes such as C-Flex and PharMed (both from Saint Gobain,) are particularly suitable for aseptic biopharmaceutical applications because of moldability, being free of animal components, and sterilizability (while thermoplastic, which makes sealing and welding easy). C-Flex is a unique, patented thermoplastic elastomer specifically designed to meet the critical demands of the medical, pharmaceutical, research, biotech, and diagnostics industries.

    C-Flex biopharmaceutical tubing has been used by many of the world's leading biotechnological and pharmaceutical processing companies for over 20 years. Each coil of C-Flex tubing is extruded to precise ID, OD, and wall dimensions. All tubing is formulated to meet the standards of the biopharmaceutical industry and is QA tested before leaving the production facility.

Features/Benefits
- Complies with USP 24/NF19, Class VI, FDA and USDA standards
- Manufactured under strict GMPs
- Nonpyrogenic, noncytotoxic, nonhemolytic
- Chemically resistant to concentrated acids and alkalies
- Significantly less permeable than silicone
- Low platelet adhesion and protein binding
- Ultrasmooth inner bore
- Superior to PVC for many applications, with significantly fewer TOC extractables
- Longer peristaltic pump life
- Heat-sealable, bondable, and formable
- Remains flexible from $-50°F$ to $275°F$
- Sterilizable by radiation, ETO, autoclave, or chemicals
- Available in animal-derived component free (ADCF), clear, and opaque formulations
- Lot traceable
- Safer disposal through incineration

Typical Applications
- Cell culture media and fermentation
- Diagnostic equipment
- Pharmaceutical, vaccine, and botanical product production
- Pinch valves

- High-purity water
- Reagent dispensing
- Medical fluid/drug delivery
- Dialysis and cardiac bypass
- Peristaltic pump segments
- Sterile filling and dispensing systems

PharMed® BPT biocompatible tubing is ideal for use in peristaltic pumps and cell cultures. PharMed® BPT tubing is less permeable to gases and vapors than silicone tubing and is ideal for protecting sensitive cell cultures, fermentation, synthesis, separation, purification, and process monitoring and control systems. PharMed® BPT tubing has been formulated to withstand the rigors of peristaltic pumping action while providing the biocompatible fluid surface required in sensitive applications. With its superior flex life characteristics, PharMed® tubing simplifies manufacturing processes by reducing production downtime due to pump tubing failure. The excellent wear properties of PharMed® BPT translate to reduced erosion of interior tubing walls, improving overall efficiency of filtering systems.

Features/Benefits
- Outlasts silicone tubing in peristaltic pumps by up to 30 times
- Low particulate spallation
- Autoclavable and sterilizable
- Temperature resistant from −60°F to 275°F
- Withstands repeated CIP and SIP cleaning and sterilization
- Meets USP Class VI and FDA criteria

Typical Applications
- Diagnostic test product manufacturing
- Cell harvest and media process systems
- Vaccine manufacturing
- Bioreactor process lines
- Production filtration and fermentation
- Sterile filling
- Shear-sensitive fluid transfer

Platinum-cured silicon: PureFit, SMP/SBP/SVP, Tygon 3350-3370, APST; biocompatible, no leachable additives, economical

Peroxide-cured silicon: Versilic SPX; biocompatible, no leachables

Modified polyolefin: Tygon LFL (www.tygon.com); chemically resistant, flexible, long-lasting

Modified polyvinyl chloride: Tygon LFL, chemical resistant, long-lasting

There are new products introduced routinely, and the reader is recommended to refer to current information on these products. One of the best
sources to meet just about all needs for tubing is Saint-Gobain; one ought to
consult with them first.

Fittings and Accessories

Connections between bags or other process stages are done by fittings
that come in a wide range of configurations, materials, and sterility. This
includes straight couplers, Y-couplers, T-couplers, cross couplers, elbow
couplers, and barbed plugs. This is necessary to allow ready solutions
to go through the often complex routines of liquids in a bioprocessing
facility. The size of these connectors ranges from $1/16$ to 1 inch in most
instances; often-incompatible sizes are downgraded or upgraded by
interim connectors called reduction couplers that are available for most
types of connectors.

The barbed plug is the most convenient as it can be easily patented with
ties to provide a very secure connection.

The tube-to-tube fittings can serve to change the size and are available in
a variety of connection options. Also available are caps to close the tube end
with the connector attached to transport the components.

LuerLok	Male and female parts are connected securely via a thread; suitable for small-volume flow rates (hose barb: 1,116-3,116 in.).
Sanitary fittings	Also known as tri-clamps (TC) genderless; a clamp connects both parts and secures a gasket between them. A connection with conventional sanitary fittings made of stainless steel is also possible (hose barb: 1/4–1 in.).
Quick (dis)connect fittings	Male and female parts are connected securely via a click mechanism. An O-ring fitted to the male part provides the seal. Pressing a button on the female part breaks the connection (hose barb: 3/8–1.5 in. also with sanitary termination).

Clamps are used for blocking or regulating flows and come in a variety of
types, the most common being the inch clamp for quick starting and stop-
ping flow; ratchet clamps adjust the flow rates. Special clamps with mechanical power transmission such as from Biovalves (www.biouretech.com),

which maintain the contact pressure via a thread arbor, are available for larger tubes with thicker walls.

The BioPure BioValves is a precision restriction flow controller and shut-off valve for the silicone tube for use in bioprocessing and pharmaceutical manufacturing applications. It is profiled to minimize flow path turbulence and can be used one-handed. Its thread pitch is calibrated to 2 mm per turn, permitting accurate estimation of flow restriction. It is molded from glass-reinforced Nylon USP Class VI. These can be repeatedly autoclaved at 134°C for 5 min or irradiated at 60 kGy (6 Mrad) with no detectable weakening.

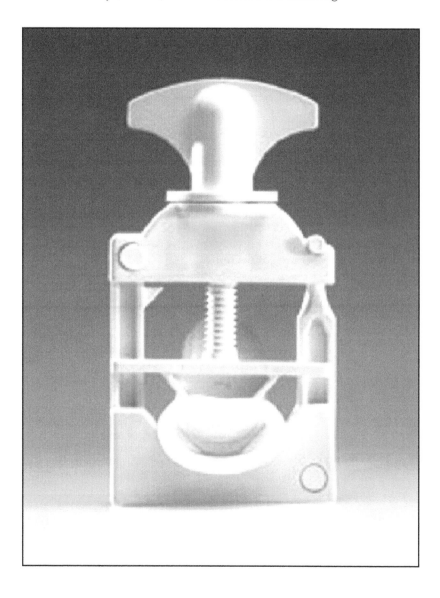

The aseptic connectors once installed cannot be disconnected to maintain patency of the process.

Pumps

Pumps are used for fluid transfer by creating hydrostatic pressure or by differential pressure; the maximum allowed pressure would be determined by the weakest part of the bioprocess component exposed to the pressure. Peristaltic pumps, syringe pumps, and diaphragm pumps are all currently used to provide disposable pumping solutions. All of these are volume displacement pumps, are easy to use, and avoid contact with the product; they can, however, produce stress on the tubing especially when the operations are conducted for an extended period of time. It is for this reason that special peristaltic pump tubes are made available by Saint-Gobain. The stress on the tube may produce particles from erosion of the tube and contaminate the fluids being passed through.

High-end peristaltic dispensing pumps have benefited from improved pulsation-free pump head design, a precise drive motor, and a state-of-the-art calibration algorithm. They are exceptionally accurate at microliter fill volumes. Peristaltic pumps that incorporate disposable tubing eliminate cross-contamination and do not require cleaning because the tubing is the only part that comes into contact with the product. Similarly, the cleaning validation of peristaltic pumps with disposable tubing is significantly easier than for piston pumps. The cost of labor and supplies for writing and executing protocols, cleaning, and documenting the cleaning process is higher for a multiple-use piston-pump filling system. Adjusting the flow speed, and therefore preventing foaming or splashing, is easier for a peristaltic pump than for a piston pump. Operators can also use a ramp-up and ramp-down feature to determine how fast a peristaltic pump reaches its fill speed. This option helps optimize overall fill time and increase throughput.

Many biological drugs are shear-sensitive, and peristaltic pumps protect them by applying low pressure and providing gentle handling. In contrast, a piston pump's valve system generates fast flow through small orifices, potentially damaging biological products. Even valveless piston pumps apply high pressures and high shear factors that could harm a biological product.

On the other hand, viscous products can be problematic for peristaltic pumps. The pumps apply only approximately 1.3 bar of pressure, and their accuracy suffers when they handle products more viscous than 100 cP.

A diaphragm pump is a positive displacement pump that uses a combination of the reciprocating action of a rubber, thermoplastic, or Teflon diaphragm, and suitable nonreturn check valves to pump a fluid. Sometimes, this type of pump is also called a *membrane pump*. There are three main types of diaphragm pumps:

- Those in which the diaphragm is sealed with one side in the fluid to be pumped, and the other in air or hydraulic fluid. The diaphragm is flexed, causing the volume of the pump chamber to increase and decrease. A pair of nonreturn check valves prevents reverse flow of the fluid.
- Those employing volumetric positive displacement where the prime mover of the diaphragm is electromechanical, working through a crank or geared motor drive. This method flexes the diaphragm through simple mechanical action, and one side of the diaphragm is open to air.
- Those employing one or more unsealed diaphragms with the fluid to be pumped on both sides. The diaphragms again are flexed, causing the volume to change.

When the volume of a chamber of either type of pump is increased (the diaphragm moving up), the pressure decreases, and fluid is drawn into the chamber. When the chamber pressure later increases from decreased volume (the diaphragm moving down), the fluid previously drawn in is forced out. Finally, the diaphragm moving up once again draws fluid into the chamber, completing the cycle. This action is similar to that of the cylinder in an internal combustion engine.

Diaphragm pumps have good suction lift characteristics, some are low pressure pumps with low flow rates, others are capable of higher flows rates, dependent on the effective working diameter of the diaphragm and its stroke length. They can handle sludges and slurries with a relatively high amount of grit and solid content. They are suitable for discharge pressure up to 1,200 bar and have good dry running characteristics. Similar to peristaltic pumps, they are low-shear pumps and can handle highly viscous liquids.

Mini diaphragm pumps operate using two opposing floating discs with seats that respond to the diaphragm motion. This process results in a quiet and reliable pumping action. Higher efficiency of the pump is evident in the longer life of the motor pump unit. These DC motor diaphragm pumps have an excellent self-priming capability and can be run dry without damage, rated to 160°F (70°C). No metal parts come in contact with materials being pumped; diaphragms and check valves are available in Viton, Santoprene, or Buna-N construction. So, these mini diaphragm pumps are very chemically resistant. The mini diaphragm pumps prime within seconds of turning the pump on; prime is maintained by two check valves (one on either side). Separated from the motor, the pump body contains no machinery parts, so the pump can be in dry running condition for a short while. A built-in pressure switch inside the pump can automatically stop the pump when the pressure reaches a specified level.

The disposable diaphragm pump head must be integrated into the transfer line prior to sterilization. As the pump head is totally closed, no other part of the pump comes into contact with the fluid. After the process, the pump head is disposed of, together with the rest of the transfer line. Flow rates of 0.1–4,000 L/h can be achieved with disposable diaphragm pumps such as from Quattroflow (www.quattrowflow.com).

Aseptic Coupling

One of the most commonly used method is to connect the tubes or components using sterile connectors under a laminar flow hood; however, this is not always possible specially when the components such as disposable bags are large and cannot be moved.

Some connectors require installation in a laminar hood followed by sterilization. These are called *SIP connectors*. Two aseptic systems go through sealing using these connectors following sterilization by autoclave, radiation, or chemical treatment. Examples of these SIP connectors are from Coler (www.coler.com) and EMD Millipore (www.millipore). The Lynx ST system from EMD Millipore comprises an integrated valve, which can be opened and closed after sterilization of the connection.

Aseptic Connectors

Critical to effective disposable processing operations are aseptic connection devices. Pharmaceutical manufacturers typically make about 25,000 aseptic connections each year, with some large manufacturers making as many as 100,000 aseptic connections annually.

The most convenient connectors are aseptic connectors that allow aseptic connections in an open uncontrolled environment without using a laminar flow hood. Examples of these aseptic connectors include the offering from Pall, Sartorius-Stedim, GE Healthcare, EMD Millipore, and Saint-Gobain. The aseptic parts on the connector side are sealed with sterile membrane filters or caps. After coupling, the sterile membrane filters must be withdrawn, and both parts have to be clamped or fixed. These connectors are secure and recommended to save time but offer an expensive choice and at times there is a limitation of sizes of tubes that can be connected. These connectors are also used as aseptic ports in bioreactors.

One of earliest entries in the field of aseptic filters was Pall Kleenpak Sterile Connector.

AseptiQuik™ Connectors (www.colder.com) provide quick and easy sterile connections, even in nonsterile environments. AseptiQuik's "click-pull-twist" design enables users to transfer media easily with less risk of operator error. The connector's robust design provides reliable performance without the need for clamps, fixtures, or tube welders. Biopharmaceutical manufacturers can make sterile connections with the quality and market availability they expect from the leader in disposable connection technology.

The Opta® SFT Sterile Connector by Sartorius-Stedim (www.sartorius-stedim.com) is a disposable device, composed of presterilized female and male coupling body, that allows a sterile connection in biopharmaceutical manufacturing processes. The Opta® SFT-I Connector is supplied with Flexel® 3D, Flexboy® bags, and transfer sets as part of integrated Sartorius Stedim Biotech Fluid Management assemblies. Opta SFT-I is available with a $1/_4$, $3/_8$, and $1/_2$ in. hose barb. The Opta® SFT-D is available as individual device for end-user assembly with TPE tubing and autoclave sterilization. They are quick, easy to use, and are backed by extensive validation work as well as 100% in-house integrity testing.

Pall Corporation (www.pall.com) is expanding its line of Kleenpak™ Aseptic Connectors with two new sizes: $1/_4$ and $3/_8$ in. The new sizes enable vaccine manufacturers to apply the safety and efficiency benefits of instant aseptic connections throughout more of their disposable operations to help speed time to market and comply with GMPs. Pall revolutionized the aseptic connection process by shortening the time needed for connection from 15 minutes to seconds when it introduced its $1/_2$-in. Kleenpak Connector. The addition of the two new Kleenpak Connector sizes increases flexibility to implement aseptic connections in more applications to improve disposable processing efficiency. This is especially important to complex vaccine production, which often requires a greater number of connection steps. The Kleenpak Connector is easy-to-use and projects an audible snap to signify that a sterile connection has been established.

ReadyMate Disposable Aseptic Connector (DAC) from GE Healthcare (www.gelifesciences.com) provides connections for high-fluid throughput and offers a secure, simple, and economical connection for upstream and downstream applications. DAC connectors can be autoclaved or gamma radiated, and can be part of a sterile circuit. The connectors can be used to connect unit operations and assemblies. DAC connectors and their components are manufactured in compliance with the cGMPs of the FDA and ISO 9000-2000. ReadyMate is a genderless, intersize connectable disposable aseptic connector. There are four hose barb sizes ($3/_4$, $1/_2$, $3/_8$, and $1/_4$ in.), mini TC, and TC that all interphase. It has a genderless design, user-friendly sanitary coupling, easy to use with Tip'n'latch, complies with USP Class VI, and can be sterilized by radiation or autoclave. The main advantages of using the Bio Quate connector include simple setup, rapid connection, direct connection of different tube sizes, direct connection of different tube materials, aseptic on-site manifold fabrication, large and smooth inside bore, no capital equipment to purchase, and no requirement for power or calibration or service. These connectors are supplied by Bioquate (www.bioquate.com).

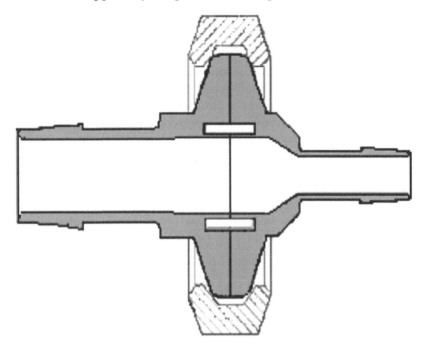

Welding

When it is possible to use a thermoplastic tube, welding offers an easy, inexpensive, and very secure solution. Examples of thermoplastic tubes include C-Flex, PharMed, and Bioprene. Both thermoplastic tubes must be aseptic, should have the same dimensions (inner diameter and OD), and should have their ends capped. The tubes are placed parallel in opposite directions, while a heated blade cuts through them and seals them simultaneously. Preheating of the blade is necessary both to achieve the welding temperature and to sterilize and depyrogenize the blade itself prior to the welding process. The depyrogenize procedure normally lasts 30 s at 250°C or 3 s at 320°C. After being cut, the tubes are moved against each other so that the ends of each tube, which are connected to the aseptic systems, are positioned directly opposite to each other on either side of the blade. A welding cycle can be between 1 and 4 min, depending on the material and the diameter of the tubes. The main welding systems available today include Sterile Tube Fuser (GE Healthcare), BioWelder (Sartorius-Stedim), Aseptic Sterile Welder 3960 (SEBRA, www.Sebra.com), TSCD (Terumo, www.terumotransfusion.com), and SCD 11B (Terumo). (Terumo supplies its equipment mainly to the blood transfusion industry.) Both GE Healthcare and Sartorius-Stedim lead the installations in the bioprocessing industry.

The Hot Lips Tube Sealer by GE Healthcare is a portable device used to thermally seal thermoplastic tubing for the transport and setup of inoculums, culture, media, and buffers. The seal forms a tamperproof and leakproof closure. Preprogrammed for a wide range of tubing types, diameters, and wall thicknesses, a single button initiates the sealing operation. The instrument is self-calibrating, and a microprocessor-controlled motor ensures repeatable performance without the need for tubing adaptors.

The Sterile Tube Fuser also from GE Healthcare is an automated device for welding together a wide range of tubing types intended for aseptic operation. Operated via a single push-button operator interface, it connects tubing between sterile containers, Cellbag bioreactors, and process equipment for the aseptic transfer of large volumes of fluids such as inoculum, media, buffers, and process intermediates.

Aseptic Transfer Systems

Moving the product across clean rooms may involve, at times, long distances; while transfer tubings between upstream and downstream areas and passthrough autoclaves are common, larger volumes transfer systems are offered by Sartorius-Stedim, ATMI Life Sciences, Getinge, and LaCalhene, which essentially constitute double-door systems using disposable containers. In using these systems, the main reusable port is always permanently fixed in the separating wall (in a clean room or isolator) and represents the containment barrier. The second connecting part is an integral part of the disposable container that stores or conducts the components, fluids, and powders to be transferred. Both the connecting parts and the reusable containers and transfer systems can be coupled to the main port. After coupling, the ports are opened from inside the cleaner area and the transfer is started. The disposable container is normally the package for the fluid and the sterile barrier for the fluid conduction.

Biosafe® Aseptic Transfer Equipment: The Biosafe® range of aseptic transfer ports offers reliable and easy-to-use solutions for the secure transfer of components, fluids, and powders while maintaining the integrity of the critical areas (such as isolators, Restricted Access Barriers (RABS), and clean rooms).

Biosafe® Aseptic Transfer Disposable Bag: A complete range of Biosafe Aseptic Transfer Bags, either gamma sterile or autoclavable, and the Biosafe® Rapid Aseptic Fluid Transfer (RAFT) System are designed to best fit requirements for aseptic transfer of components into clean rooms, isolators, or RABS and for contained transfer of potent powders.

Biosafe® RAFT System: The RAFT system provides easy-to-use and reliable through-the-wall aseptic transfer of liquid between clean rooms of different environmental classification while ensuring a total confinement.

SART™ System: The SART System is designed to allow aseptic liquid transfer between two areas with different containment classifications.

Special bags have therefore been developed, for example, the Biosafe Rapid Aseptic Fluid Transfer (RAFT) system by Sartorius-Stedim, allowing aseptic coupling to larger fluid containers and ports in addition to fluid conduction.

Tube Sealers

When disconnecting an aseptic connection, the ends must be capped with aseptic caps and this can be done under a laminar hood or by using tube sealers; the examples of which include offerings from PDC (www.pdcbiz.com), Saint-Gobain (www.saint-gobain.com), Sartorius-Stedim (www.sartorius-stedim.com), GE Healthcare (www.gelifesciences.com), Terumo (www.terumotransfusion.com), and SEBRA (www.sebra.com). Most of these sealers can seal from 0.25 to about 1.5 in. tubes and take from 1–4 min to complete the seal. Most operate on an electrical heating element, but electrical and radio frequencies are also used for sealing tubes. There is no need for using a laminar flow hood for these operations. In most instances, applying a crimper in two places and cutting the tube between the crimps offers the cheapest solution.

Sampling

Sampling is a routine during manufacturing to ensure compliance by obtaining these in process parameters such as pH, DO, OD, pCO2, etc. Most disposable systems have one or more integrated sampling lines, which are partly equipped with special sampling valves, sampling manifolds, or special sampling systems. A popular disposable sample valve is the Clave connector from ICU-Medical (www.icumed.com), which is also used in intravascular catheters for medical applications. It allows a sample to be taken with a

LuerLok syringe. A dynamic seal inside the valve guarantees that the sample is not taken until the syringe is connected, thereby ensuring the sample only comes in contact with the inner, aseptic parts of the valve. However, the samples drawn do not remain sterile.

Manifolds consisting of sampling bags, sampling flasks, or syringes are appropriate for taking aseptic samples in disposable systems. These manifolds can be connected to the systems via aseptic connectors or tube welding. Sampling manifolds allow multiple sampling for quality purposes over a given period of time. The main feature of the manifold is that the number of manipulations in a process is significantly reduced. The manifold systems are delivered ready for process use, preassembled, and sterile. Only one connection has to be made to allow several bags to be filled.

Also used for sampling are manifold systems where sample containers of a manifold are arranged in parallel whereby the last one is used as a waste container. Through using Y-, T-, or X-hose barbs and tube clamps, the initial flow and the subsequent sample are guided to the appropriate containers. SIP connections, of course, also allow the connection of manifold systems to conventional stainless steel processing equipment.

Conclusion

The complexity of bioprocessing makes it difficult to design systems without any weak links; contamination is indeed the most significant risk, which requires that all connectors, tubing, and implements joining various steps of a process and performing sampling remain patent. Disposable connectors and tubing were one of the first components that went disposable. Still, in hard-walled systems, SIP systems are in use only because there is steam for CIP/SIP operations. Even then, the risk of contamination remains. Since much of the disposable technology in these applications has come from the biomedical field, the device industry had always been ahead of the regulatory requirements. Biocompatibility issues have long been resolved and vendors are able to provide detailed information on their devices that might be needed by regulatory agencies. Since the manufacturing of these devices is complex, it is unlikely for a user to request custom devices; however, the diversity of choices available today is enough to modify any system that would be able to use an off-the-shelf item. As before, the emphasis on the importance of an off-the-shelf item over custom designs remains.

The tube connectors and sealers are a newer entry as disposable bags for mixing, and bioreactors have becoming more popular; still, there is a limited choice of suppliers, mainly GE and Sartorius-Stedim. The cost of this equipment is still high, but then the alternative comes down to using expensive aseptic connectors. Generally, if a good choice of aseptic connectors is available, that should be preferred over tube connectors since it is always possible to make a poor connection using the heat-activated systems; also, the use of aseptic connectors allows connecting tubes that may not be thermolabile.

7

Controls

The goal of PAT is to understand and control the manufacturing process, which is consistent with our current drug quality system: *quality cannot be tested into products; it should be built-in or should be by design.*

Food and Drug Administration, 2009

According to the Food and Drug Administration (FDA) Guidance for Industry, process analytical technology (PAT) is intended to support innovation and efficiency in pharmaceutical development. PAT is a system for designing, analyzing, and controlling manufacturing through timely measurements (i.e., during processing) of critical quality and performance attributes of raw and in-process materials and processes with the goal of ensuring final product quality. It is important to note that the term *analytical* in PAT is viewed broadly to include chemical, physical, microbiological, mathematical, and risk analyses conducted in an integrated manner.

To fulfill process requirements, single-use sensors, which are either integrated in the single-use bioreactor or included in the cover and are disposed of with the bioreactor, are required. They provide a continuous signal and allow information about the status of the cell culture to be gathered at any time. The traditional batch analysis such as high-pressure liquid chromatography (HPLC), electrochemistry, and wet chemical analysis in place of disposable sensors increases the risk of contamination.

Since disposable bioreactors are new to the industry, the first attempt to monitor the product in the bioreactor was to use the traditional biosensors used in hard-walled systems to measure bioreactor temperature, dissolved oxygen (DO), pH, conductivity, and osmolality. These probes must first be sterilized (via autoclaving) and then attached to penetration adapter fittings that are welded into bioreactor bags. Not surprisingly, this is a labor-intensive and time-consuming process that has the potential to compromise the integrity and sterility of single-use bioreactor bags, and has been largely discarded in favor of truly disposable sensors. Critical process parameters that are often monitored include pressure, pH, DO, conductivity, UV absorbance, flow, and turbidity. The packages that contain the traditional technologies for monitoring these parameters are not usually compatible with or effective when integrated into single-use assemblies for many reasons: cost, cross-contamination, inability to maintain a closed system, and system incompatibility with gamma irradiation.

The practice of integrating bags, tubing, and filters into preassembled, ready-to-use bioprocess solutions is optimized if noninvasive sensing of critical process parameters is part of the package instead of using sensors that may require sterilization and cleaning validation, the core processes which are obviated in the use of disposable bioreactors.

Even though these obstacles do not always preclude the use of traditional measurement technologies, single-use solutions for monitoring process parameters eliminate the need for equipment cleaning and autoclaving small parts, reduce the risk and cost involved with making process connections, and may be more cost-effective than tracking and maintaining traditional technologies. For example, a sanitary, autoclavable pressure transducer that is qualified for a certain number of autoclave cycles and requires recalibration may be more expensive to use versus a single-use pressure sensor.

The adoption of disposable sensors requires a keen understanding of their need and utilization. Their suitability would be determined by their material properties, sensor manufacturing, process compatibility, performance requirements, control system integration, compatibility with treatments before use, and regulatory requirements.

Several companies, including Finesse and Fluorometrix (recently acquired by Sartorius-Stedim), have created single-use, membrane biosensors that can be added to or directly incorporated (during manufacturing) into single-use bioreactor bags.

There are two options in using disposable sensors: one where the sensors are placed in situ in contact with the liquid, and the other where the external sensors contact the medium either optically (ex situ) or via a sterile (and disposable) sample removal system (on line). Disposable sensors must be sterilizable if they come in contact with media; these must also be cost-effective and reliable. Better designs using inexpensive sensing elements can be located inside a disposable bioreactor and combined with reusable (and more expensive) analytical equipment outside the reactor. Inexpensive, single-use sensors can also be placed on transistors and placed either in the headspace, inlet, outlet, or into the cultivation broth for liquid-phase analysis (temperature, pH, pO_2). These can also be optical sensors that allow noninvasive monitoring through a transparent window.

Sampling Systems

Continuous sampling from a bioreactor can be accomplished using a sterile filter and a peristaltic pump to obtain a cell-free sample and, where the dead volume of a sample is of concern (as in smaller bioreactors), microfiltration membranes can be used that may be placed inside the bioreactor; disposable

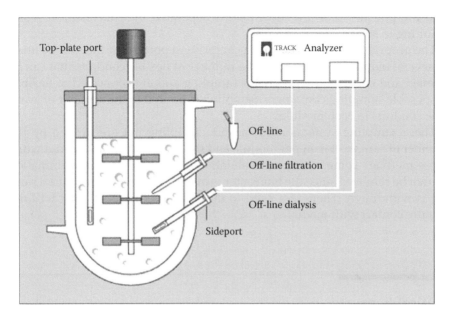

FIGURE 7.1
Groton BioSystems Sampling Method.

forms of these are not yet available. Suppliers include TraceBiotech (www. trace.de) and Groton (www.grotonbiosystems.com) (Figure 7.1).

TRACE System

Where removing cells prove cumbersome, the samples may be treated to stop their metabolic activity by freezing or using inactivation chemicals.

One way to solve the sampling problem is to use a presterilized sampling container, including a needleless syringe that can be welded to the sampling module of the bag bioreactor. A sample is pumped into the container, the sampling containers can be removed, and the tube heat-sealed. Sartorius and GE use this method. Other fully sampling systems involve connecting to a bioreactor a presterilized Luer connection including a one-way valve to prevent the sample from flowing back into the reactor. The sample is withdrawn from the reactor by a syringe and directed through a sample line into a reservoir. Cellexus Biosystems (Cambridgeshire, UK) and Millipore (Billerica, MA) use this approach.

The Cellexus system is connected to the sample line, and there are up to six sealed sample pouches. The sample from the reservoir can then be pushed into the pouches that are subsequently separated by a mechanical sealer resulting in sealed, sterile samples.

The proprietary Millipore system comprises a port insert that can be fitted to several bioreactor sideports and a number of flexible conduits that can be opened and closed individually for sampling and are connected to flexible, disposable sampling containers. Sampling is limited to the number of available conduits in each module.

These sampling systems allow aseptic sampling but are limited by the number of samples taken per module and the lack of automation. And while these methods come with good validation data, the risk of contamination cannot be removed since the bioreactor is indeed breached every time a sample is withdrawn. There is a need to develop other methods that will not require contact with media.

Optical Sensors

Optical sensors work on the principle of the effect of electromagnetic waves on molecules. It is an entirely noninvasive method and can provide continuous results of many parameters at the same time. It is relatively easy to use them through a transparent window in the bioreactors. The detector part of the system can be physically separated, allowing utilization of expensive analytical devices allowing optical sensors to be used in situ or online.

Fluorescence sensors can be optimized for measurements of nicotinamide adenine dinucleotide phosphate (NADPH) and are used for both biomass estimation and differentiating between aerobic and anaerobic metabolism. The two-dimensional process fluorometry enables simultaneous measurement of several analytes by scanning through a range of excitation and emission wavelengths including proteins, vitamins, coenzymes, biomass, glucose, and metabolites such as ethanol, adenosine-5′-triphosphate (ATP), and pyruvate. Thus, it is possible to use fluorometry to characterize the fermentation process. Generally, a fiber-optic light attached to the bioprocessor and shining the light through a glass window in the bioprocessor works very well. An example of this is the fluorometers from BioView system (www.delta. dk). The BioView sensor is a multichannel fluorescence detection system for application in the biotechnology, pharmaceutical and chemical industries, food production, and environmental monitoring. It detects specific compounds and the state of microorganisms as well as their chemical environment without interfering with the sample. The BioView system measures fluorescence online directly in the process. An interference with the sample is eliminated. There is no need to take samples for off-line analysis that saves manpower and reduces the risk of contamination. However, in view of the complexity of spectra of multiple components, high-level resolution programming is required.

Finesse Solutions, LLC (www.fiesse.com), a manufacturer of measurement and control solutions for life sciences process applications, announced a live demonstration of its new SmartBag product for rocker bioreactors at Interphex 2011 in New York, March 29–31. SmartBags are designed to be plug-and-play bioprocessing containers having full measurement capability for at least 21 days.

The SmartBag SensorPak leverages TruFluor pH and DO phase fluorometric technology in a compact assembly that is precalibrated using a SmartChip and provides accurate, drift-free, in situ measurements. The combined pH and DO optical reader uses advanced optical components including a large area photodiode that minimizes photodegradation of the active-sensing elements. The SensorPak also leverages TruFluor temperature 316 L stainless steel thermal window for highly stable readings. The SensorPak is welded into the single-use vessel and eliminates the need for sterile connectors and their associated complications such as leakage and batch contamination. All wetted materials of the SensorPak are USP class VI compliant and, being identical to TruFluor, allow directly measurement comparisons and scale-up from 10 L rocker bags to 2,000 L SUBs.

The biosensors manufactured by Fluorometrix are noninvasive, membrane sensors developed using optical fluorometric chemistries that can be directly incorporated into any disposable bioreactor bags. Because the sensors can be manufactured into any type of single-use bag, they are useful for both upstream and downstream applications. Also, these are compatible with the FDA's PAT initiative.

Many metabolic products in a bioreactor can be readily detected by IR spectroscopy but water-absorbed IR beam can only be NIR or SIR for biomass analysis when used in the transmission mode. However, attenuated total reflectance spectroscopy (ATRIR) is based on the reflection of light at an interface of two phases with different indices of refraction, and the light beam penetrates into the medium with the lower refraction index in the dimension of one wavelength. Absorption of IR results in decrease in the intensity of the reflected beam to detect the analyte. Probes for both types of IR spectroscopy are used. Hitec-Zang (www.hitec-zang.de) offers a large range of PAT devices including IR systems.

NIR transmission probes and ATR IR probes for bioreactors are now commercially available. These are connected through silver halide fibers or radio frequency connectors.

In addition to IR and fluorescence, optical methods based on photoluminescence, reflection, and absorption are also used. The optical electrodes or "optodes" can be attached using glass fibers leaving the measurement equipment outside of the bioreactor as discussed earlier for fluorescence detectors allowing use of these chemosensors in situ or online.

Oxygen sensors work by quenching fluorescence by molecular oxygen; measurement requires a fluorescent dye (metal complexes) immobilized and attached to one end of an optical fiber, and the other end of the fiber is interfaced with an excitation light source. The duration and strength of fluorescence depend on the oxygen concentration in the environment around the dye. The emitted fluorescence light is collected and transmitted for reading outside of the bioreactor. These electrodes work better than the traditional platinum probe electrodes to detect oxygen, working in both liquid and gas phases. Examples of oxygen sensors include PreSens (www.presens.de) noninvasive oxygen sensors that measure the partial pressure of both dissolved and gaseous oxygen. These sensor spots are used for glassware and disposables. The sensor spots are fixed on the inner surface of the glass or transparent plastic material. The oxygen concentration can therefore be measured in a noninvasive and nondestructive manner from outside, through the wall of the vessel. Different coatings for different concentration ranges are available. It offers online monitoring of concentration ranges from 1 ppb up to 45 ppm DO, with dependence on flow velocity and measuring oxygen in the gas phase as well; these can be autoclaved.

Ocean Optics (www.oceanoptics.com) offers the world's first miniature spectrometer with a wide array of sensors for oxygen, pH, and in the gas phase.

The pH sensors work by fluorescence or absorption, and for fiber-optic pH measurements, both fluorescence- and absorbance-based pH indicators can be applied. For fluorescence, the most common dyes are 8-hydroxy1,3,6-pyrene trisulfonic acid and fluorescein derivatives, while phenol red and cresol red are used for absorption-type measurements. Fluorescent dyes are sensitive to ionic strength limiting their use for broad pH measurement, more than 3 units.

The new transmissive pH probes from Ocean Optics use a proprietary sol–gel formulation infused with a colorimetric pH indicator dye. This material is coated onto the exclusive patches to reflect light back through the central red fiber or to transmit light through in order to sense the color change of the patch at a specific wavelength. While typical optical pH sensors are susceptible to drastic changes in performance in various ionic strength solutions, Ocean Optics' sensory layer has been chemically modified (esterified) to allow accurate sensing in both high- and low-salinity samples. The transmissive pH probes from Ocean Optics can be used with a desktop system as well as with the Jaz handheld spectrometer suite. The desktop system uses a module is SpectraSuite software that allows for simplified calibration, convenient pH readings, customizable data logging, and comprehensive exportation of data and calibration information.

- Proprietary organically modified sol–gel formulation engineered to maximize immunity to ionic strength sensitivity.

- Compatible with some organic solvents (acetone, alcohols, aromatics, etc.).

- Sol–gel material chosen over typical polymer method, allowing for faster response time, versatility in the desired dopants, greater chemical compatibility, flexible coating, and enhanced thermal and optical performance.

- Indicator molecule allows high-resolution measurement in biological range (pH 5–9).

- Simplified algorithm takes analytical and baseline wavelengths into account to reduce errors caused by optical shifts.

The TruFluor™ (www.finesse.com) DO and temperature sensor is a single-use solution consisting of a disposable sheath, an optical reader, and a transmitter. The single-use sheath can be preinserted in a disposable bioreactor bag port and irradiated with the bag to both preserve and guarantee the sterile barrier. All wetted materials of the sheath are USP class VI compliant. The optical reader utilizes an light emitting diode (LED) and a large area photodiode with integrated optical filtering that minimizes photodegradation of the acting sensing element. The design has been optimized to provide accurate in situ measurement of dissolved oxygen using phase fluorometric detection in real time. The temperature measurement leverages a 316 L stainless steel thermal window embedded in the sheath and provides a highly accurate temperature measurement that can be used as a process variable or for temperature compensation.

Carbon dioxide sensors work on the principle of measure pH of a carbonate buffer embedded in a CO_2-permeable membrane. The reaction time of the sensors is long, and the use of quaternary ammonium hydroxide has been made to achieve a faster response. Fluorescence-based sensors are attractive as they facilitate the development of portable and low-cost systems that can be easily deployed outside the laboratory environment. The sensor developed for this work exploits a pH fluorescent dye 1-hydroxy-pyrene-3,6,8-trisulfonic acid, ion-paired with cetyltrimethylammonium bromide (HPTS-IP), which has been entrapped in a hybrid sol–gel-based matrix derived from *n*-propyltriethoxysilane along with the lipophilic organic base. The probe design involves the use of dual-LED excitation in order to facilitate ratiometric operation and uses a silicon P-type intrinsic n-type (PIN) photodiode. HPTS-IP exhibits two pH-dependent changes in excitation bands, which allows for dual-excitation ratiometric detection as an indirect measure of the pCO2. Such measurements are insensitive to changes in dye concentration, leaching, and photobleaching of the fluorophore and instrument fluctuations unlike unreferenced fluorescence intensity measurements. The performance of the sensor system is characterized by a high degree of repeatability, reversibility, and stability.

The YSI 8500 CO_2 monitor measures dissolved carbon dioxide in bioprocess development applications. Engineered to fit within a variety of bioreactors, the unit delivers precise, real-time data that increase an understanding of critical fermentation and cell culture processes. This data can help in gaining insight into cell metabolism, cell culture productivity, and other changes within bioreactors.

An in situ monitor based on the reliable optochemical technology was developed by Tufts University and YSI Incorporated (www.ysilifesciences. com). The technology involves the use of a CO_2 sensor capsule consisting of a small reservoir of bicarbonate buffer covered by a gas-permeable silicone membrane. The buffer contains hydroxypyrene trisulfonic acid (HPTS), a pH-sensitive fluorescent dye. CO_2 diffuses through the membrane into the buffer, changing its pH. As the pH changes, the fluorescence of the dye changes. The model 8500 monitor compares the fluorescence of the dye at two different wavelengths to determine the CO_2 concentration of the sample medium. The sensor can be autoclaved multiple times. It will measure dissolved CO_2 over the range of 1% to 25%, with an accuracy of 5 % of the reading, or 0.2% absolute. Previously, CO_2 was measured either in the exit gases from the fermentation process or by taking a manual sample. The new optical-chemical technology uses a fiber-optic cable transfer light through a stainless steel probe into a disposable sensor capsule, which contains a pH sensitive dye. The dissolved CO_2 diffuses through a polymer membrane to change the color of the dye, which is then relayed by fiber-optic cable back to a rack-mounted monitor that determines and displays the dissolved CO_2 level.

Biomass Sensors

Information about the biomass concentration can also be obtained via turbidity sensors. Generally, these sensors are based on the principle of scattered light. Most turbidity sensors have the disadvantage that there is only a linear correlation for low particle concentrations. But sensors that use backscattering light (180°) also have linear properties for high particle concentrations. A window that is translucent for the desired wavelength in the IR region is necessary for the use in disposable reactors. The S3 Mini-Remote Futura line of biomass detectors (www.applikonbio.com) makes it possible to incorporate sensors inside disposable bioreactors. This system incorporates an ultra lightweight preamplifier for connecting to the Aber Instrument Company (ABER) disposable probe. The main Futura housing can be mounted away from the single-use bioreactor vessel. Cells with intact plasma membranes in a fermenter can be considered to act as tiny capacitors under the influence of an electric field. The nonconducting nature of the plasma membrane allows a build up of charge. The resulting capacitance can be measured: it is dependent upon the cell type and is directly proportional to the membrane-bound volume of these viable cells. The choice of in situ steam-sterilizable probes includes a single-use, sterilizable flow through the cell.

Electrochemical Sensors

Electrochemical sensors include potentiometric, conductometric, and voltametric sensors. Thick- and thin-film sensors, as well as chemically sensitive field-effect transistors (ChemFETs), possess potential as potentiometric disposable sensors in bioprocess control because they can be produced inexpensively and in large quantities.

Many pH-sensing systems rely on amperometric methods, but they require constant calibration due to instability or drift. The setups of most amperometic sensors are based on the pH-dependent selectivity of membranes or films on the electrode surface.

While turbidity sensors detect the total amount of biomass concentration, capacitance sensors provide information specifically about the viable cell mass. The electrical properties of cells in an alternating electrical field are generally characterized by an electrical capacitance and conductance. The integrity of the cell membrane exerts a significant influence on the electrical impedance, so that only viable cells can be estimated. The Biodis Series for monitoring viable biomass in disposables applications is available from Fogale (www.fogalebiotech.com) and Aber (www.aberinstruments.com), the latter now offers an integrated version with Dagsip Biotools Company (DASGIP) (www.dasgip.de). The new Aber Futura Biomass Monitor has been designed so that multiple units can easily be incorporated into bioreactor controllers and supervisory control and data acquisition (SCADA) systems.

The sensor CITSens Bio (http://www.c-cit.ch/) can monitor the consumption of glucose and/or the production of l-lactate during cultivation. The CITSens Bio utilizes an enzymatic oxidation process and electron transfer from glucose or lactate to the electrode (anode) via a chemical wiring process, which is catalyzed by an enzyme specific for ~-d-glucose or l-lactate and a mediator. The sensor function is therefore not affected by oxygen concentration and produces an exceptionally low concentration of side products, such as peroxide. The working principle of this sensor is in contrast to that of a number of well-known alternatives currently on the market, which depend on a sufficient supply of oxygen for their operation as they measure the hydrogen peroxide produced during the bioprocess. The principal feature of the CITSens Bio is a miniaturized, screen-printed electrode comprising a three-electrode system for amperometric detection of the current transmitted to the anode (working, counter, and reference electrode). This three-electrode system ensures a reliable electrical signal with long-term stability. The chemical components, including the enzyme, are deposited onto the active field of the working electrode, and the enzyme is cross-linked to form protein and hence is immobilized in this network. The immobilization process itself has an antimicrobial effect. A dialysis membrane is cast over the sensing head to create a barrier between the sensor and the cultivation medium.

Pressure Sensors

Another important process parameter that is frequently monitored during bioprocess unit operations such as filtration, chromatography, and many others is pressure. Using a traditional stainless steel pressure gauge in conjunction with a disposable experimental setup is possible, but has the drawback that the pressure gauge has to be sterilized separately. Furthermore, the connection of the sensor to the previously gamma-radiated disposable assembly can be problematic.

Many bioprocess unit operations are either controlled based on pressure or have significant pressure-related safety issues. Traditional stainless steel reactors are monitored and controlled for pressure, as pressure is used as a means of influencing mass transfer and preventing contamination. In addition, a high-pressure event is a potentially hazardous situation. Single-use bioreactor systems, on the other hand, are frequently not monitored or controlled for pressure because stainless steel pressure transducers are not compatible or cost effective when applied to disposable bioreactors. As a result, a clogged vent filter on a bioreactor can easily rupture bags, spilling the contents of the reactor and exposing the operators to unprocessed bulk.

Another application where pressure monitoring is central to process performance is depth and sterile filtration. A filter's capacity is primarily measured by either flow decay or pressure increases, although adding reusable traditional pressure transducers to a process train defeats the purpose of a single-use process setup. Depending on the process application, the product contact surface of a traditional device requires either sanitization or moist heat sterilization.

There are traditional devices that are compatible with steam in place (SIP), where only the product contact surface is exposed to steam, and even devices that can be placed in an autoclave where the entire device is exposed to steam. Many single-use process components, however, are not compatible with moist heat sterilization temperatures, so there may be a requirement for separate sterilization of the stainless steel device and possibly less than optimal connection to a presterilized disposable assembly.

Single-use pressure sensing allows for rapid changeover of product contact parts in both development applications and especially in early phase clinical manufacture. For example, single-use pressure sensors from PendoTECH were designed to enable pressure measurement with single-use assemblies that have flexible tubing as the fluid path. These single-use pressure sensors are gamma compatible (up to 50 KGy), and the fluid-path materials meet USP class VI guidelines and are also compliant with European Medicines Agency (EMEA) 410 Rev 2 guidelines.

On a single-use bioreactor, a sensor can be installed on a vent line to measure headspace pressure. Even though the sensors are qualified for use up to 75 psi, the core sensor is accurate in the low-pressure range required for a single-use bioreactor.

A better solution with respect to ease of operation and compatibility are disposable pressure sensors, which are now available on the market. The single-use sensors from PendoTECH (www.pendotech.com) can be used with tubing of various sizes (0.25 in. to 1 in. in diameter) and can be gamma radiated with tubing and bag assemblies. They are the alternative low-cost solution for use with tubing and bioprocess containers to the existing stainless steel pressure sensors on the market. Available in caustic-resistant polysulfone so they can be in-line during caustic sanitization processes. The pressure sensors can be integrated for pressure measurement and control with a PressureMAT™ System (monitor/transmitter) or PendoTECH Process Control System and depending on the number of sensors and process requirements. The data collected by these systems can be output to a personal computer (PC) or another data monitoring device. They also can be integrated into other prequalified third-party pumps and monitors (adapters for phone jacks can be made). The pressure sensors are very accurate in the pressure ranges typically used with flexible tubing and disposable process containers, and are qualified for use to 75 psi. Applications include multistage depth filtration, tangential flow filtration (TFF)/cross-flow filtration, and bioreactor pressure monitoring.

NovaSensor's NPC-100 (www.ge-mcs.com) pressure sensor is specifically designed for use in disposable medical applications. The device is compensated and calibrated per the Association for the Advancement of Medical Instrumentation (AAMI) guidelines for industry acceptability. The sensor integrates a high-performance, pressure sensor die with temperature compensation circuitry and gel protection in a small, low-cost package.

The SciPres (www.scilog.com) combines pressure-sensing capabilities and the convenience of disposability with easy setup. Each sensor is pre-programmed and barcoded with a unique ID for easy traceability and data documentation when combined with the SciLog SciDoc software. Factory calibration data is also stored on each sensor's chip for out-of-box, plug-and-play use. The SciPres comes in five different sizes to fit a variety of tubing sizes: Luer, 3/8″ barb, 1/2″ barb, 3/4″ TC (Tri-Clover), and 1.0″ TC (Tri-Clover).

The SciCon combines temperature-sensing capabilities with conductivity-sensing capabilities in a compact, disposable, single-use package at a low price point. Similar to the SciPres, each sensor is preprogrammed and barcoded with a unique ID for easy traceability and data documentation when combined with the SciLog SciDoc software, and factory calibration data is also stored on each sensor's chip for out-of-box, plug-and-play use. The SciCon comes in five different sizes to fit a variety of tubing sizes: Luer, 3/8″ barb, 1/2″ barb, 3/4″ TC (Tri-Clover), and 1.0″ TC (Tri-Clover).

Conclusions

The need to monitor the characteristics of a biomass goes across many industries and, most notably in the fermentation industry such as wine-making, a method of online and in situ monitoring of just about every function that is needed to perform a full PAT work is available. In recent years, there have been significant breakthroughs in the technologies available for monitoring, including fluorescence, dye–base pH, and oxygen measurements. While most of the sensors were initially developed for the hard-walled bioreactor industry, the disposable versions of these sensors are appearing almost every day. The basic principle is that if a sensor is placed inside the bioreactor vessel, it should be sterile and preferably disposable; the cost of throwing away a sensor has come down significantly, and it is now possible to readily monitor just about every function including cell mass, both total and live, using these disposable sensors. Alternately, many methods are available that work from outside of the bioreactor entirely, particularly those involving fluorescence and optical measurements.

It is anticipated that within the next 5 years as disposable biroeactors begin to replace large hard-walled systems, much improved systems will become available, especially those consolidating several monitoring functions into one.

8

Downstream Processing

Remember, a dead fish can float downstream, but it takes a live one to swim upstream.

W. C. Fields

The adoption of disposable components in downstream bioprocessing has been an evolutionary process with a few revolutionary peaks here and there. It started with buffer bags and devices for normal flow filtration, including virus filtration and guard filters for chromatographic columns, but gradually, more complex concepts have been introduced, including disposable devices for tangential flow filtration and chromatography in downstream processing. Today, the consensus of the industry is that while many of the upstream operations can be converted to fully disposable systems, at least some elements of downstream processing will remain traditional, and the reasons quoted for this assertion is that columns and resins will always be too expensive to throw away. Also, since columns can be of a very large size, it will be difficult to find a suitable disposable substitution.

However, as history tells, these were the same arguments presented just 15 years ago opposing the conversion of bioreactors to disposable devices. Today, downstream processing science is developing more rapidly than upstream science; more recently, the use of membrane adsorbers has been recommended for large-scale purification of antibodies. These membranes are much cheaper than classical resins.

For cell harvesting and debris removal, disposable filtration systems are available. Benefits include the ease of scale-up and the availability of presterilized filter capsules that can be integrated directly into production lines. Though this stage is generally completed by centrifugation or lenticular filtration, Millipore's Pod systems provide the first available alternative in disposable lenticular filters. This combines two distinct separation technologies in an adsorptive depth filter to enhance filter capacity and retention, while compressing multiple filtration steps into one efficient operation. Scale-up is achieved by inserting multiple pods into a holder, with formats allowing 1–5 or 5–30 pods as required. Further disposable depth filter formats include the Stax-System from Pall Life Science, encapsulated Zeta Plus from Cuno, and the L-Drum from Sartorius-Stedim.

The next step is cross-flow filtration to reduce the volume, but the build-up of debris extends the time for filtration and while this process is not a sterile process, use of disposable filter prevents the problem of cross-contamination.

For capturing and polishing, a steel column is packed with a resin (stationary phase) comprising porous beads made of a polysaccharide, mineral, or synthetic matrix conjugated to specific functional groups exploiting different separative principles. The protein mixed with other components is loaded onto the column slowly, and once it is bound to resin, the resin is eluted with appropriate pH and electrolyte solutions to separate the target protein from the mixture. The resin is cleaned and sanitized for repeated use that may involve dozens or perhaps hundreds of cycles. To overcome the time needed to pack resin and operate a column, several companies now offer columns such as GE's ReadyToProcess systems for use in AKTA machines. GE offers a wide range of resins and offers custom resins as well. These are high-performance bioprocessing columns that come prepacked, prequalified, and presanitized. Designed for seamless scalability, they deliver the same performance level as available in conventional processing columns such as AxiChrom™ and BPG™. Currently available with a range of BioProcess™ media in four different sizes—1, 2.5, 10, and 20 L—these are designed for purification of biopharmaceuticals for clinical phase I and II studies. Depending on the scale of operations, they can also be used for full-scale manufacturing, as well as for preclinical studies. The columns can be used in a wide range of chromatographic applications for separation of various compounds such as proteins, endotoxins, DNA, plasmids, vaccines, and viruses.

Atoll offers the MediaScout MaxiChrom columns (www.atoll-bio.com), which are disposable and totally incinerable; they come packed with Amberlite XAD-4 as disposable item for removal of detergents after virus inactivation during production in completely incinerable columns, and packed separation media for virus removal during production in completely incinerable columns, packed separation media for virus validation experiments during registration of a separation scheme in completely incinerable columns. MediaScout® MaxiChrom 100-X columns (X = 50–300 mm) are professionally packed with any resin or chromatography media chosen by the user, preferably with materials of particle size larger than 50 μm. They are individually flow-packed to take into account the varying compressibility of each resin. Bed heights are fixed to an accuracy of ±1 mm. MaxiChrom 100-X chromatography columns are designed for preparative applications and/or scale-up development work. The column hardware is fully incinerable, which makes the columns particularly useful for single use in biopharmaceutical production.

The Case of Monoclonal Antibodies: A GE Report

To reduce costs and shorten time to market, the use of plug-and-play technology is increasing in process development as well as in later-stage production. The following study was conducted by GE and compared with their standard XK columns, which can run all media types:

ReadyToProcess™ columns are prepacked, prequalified, and presanitized columns ready for direct use. In this study, the performance of ReadyToProcess columns prepacked with MabSelect SuRe™, Capto™ Q, and Capto adhere media was compared with small-scale XK16/40 (XK) columns packed with the same media for the purification of a monoclonal antibody (mAb) in a three-step process in parallel experiments. Yield and contaminant levels were practically identical during all steps, demonstrating the comparable performance of the column types and that the process is scalable. In addition, ReadyToProcess columns can be used for repeated runs with retained performance.

The increasing demand for mAbs as biopharmaceuticals has promoted the development of efficient processes for cell culturing, as well as for purification. Plug-and-play units make several time-consuming steps redundant, and therefore shorten time to market. Such solutions also reduce the risk of cross-contamination significantly. ReadyToProcess columns are prepacked, prequalified, and presanitized process chromatography columns, suited for purification of biopharmaceuticals (e.g., proteins, vaccines, plasmids, viruses) for clinical phase I and II studies. The columns are ready for use, and the design makes them easy to connect to chromatography systems and to dispose of after completed production. ReadyToProcess columns are available with a range of BioProcess™ media in several sizes. In this study, the performance of ReadyToProcess columns was compared with an established small-scale format. A mAb was purified from cell culture supernatant using a three-step, generally applicable process consisting of MabSelect SuRe, Capto Q, and Capto adhere. The BioProcess media in the ReadyToProcess columns are the same as those used in conventional process chromatography, thus allowing the use of a fully flexible mode in early production while keeping a conventional reuse option for later large-scale manufacturing open.

ReadyToProcess columns (2.5 l column volume [CV]) and XK16/40 columns (40 mL CV) packed with the same media and having the same bed height (20 cm) were used to compare the performance of the column types and to demonstrate scalability. The three-step purification strategy involved capture using MabSelect SuRe, an affinity medium with an alkali-tolerant protein A-derived ligand. Further, intermediate purification using ion exchange was employed with Capto Q followed by a final polishing step of the mAb with Capto adhere. The columns were connected to ÄKTAexplorer™ 100 (XK columns) and ÄKTAprocess™ (ReadyToProcess columns) chromatography systems. UNICORN™ software was used for control and evaluation. By using a platform approach, the development time and effort was kept to a minimum, and the development work was concentrated on the third, polishing step, where the multimodal anion exchanger Capto adhere was used.

The feed consisted of filtered CHO cell culture supernatant containing 2.7 mg mAb/mL. Sample volumes corresponding to 25 mg mAb/mL bed volume were applied to the XK 16/40 and RTP MabSelect SuRe 2.5 columns. Five cycles, each including cleaning-in-place (CIP) with 0.5 M NaOH, were run on each column, and the eluates were collected using an UV watch function. mAb purification at large scale typically contains a virus inactivation step at low pH after the protein A capture step, taking advantage of the low pH of the collected eluate. This step was omitted in this study. To match the buffer conditions of the equilibration buffer in the subsequent Capto Q step, the pH of the collected eluates was immediately adjusted to 7.6.

The pH-adjusted eluates from the five MabSelect SuRe runs were pooled and applied to the Capto Q column in flow-through mode. The flow through and part of the washing solution were collected and prepared for the Capto adhere step by adjusting the conductivity and pH to match the conditions of the equilibration buffer in the Capto adhere step.

All material from the Capto Q run was applied to Capto adhere in flow-through mode. The flow through and washing solution were collected.

Samples were withdrawn for analysis at each stage of the purification process. The amount of dimer and aggregates in the samples was determined by gel filtration on a Superdex™ 200 10/300 GL column. Host cell protein (HCP) concentration was determined using the CHO-CM HCP ELISA kit (CM015, Cygnus Technologies). The concentration of leached MabSelect SuRe ligand was determined by a protein A ELISA method using purified ligand for the ELISA standard curve. The analyses were not optimized for this particular feed and mAb. Three-step monoclonal antibody purification was performed in parallel at two different scales. The overall yield was 88% for both processes, achieving contaminant levels acceptable for formulation. The three-step purification process is characterized by an overall good yield, low ligand leakage from MabSelect SuRe, and efficient contaminant and dimer/aggregate removal. It should be emphasized that the process development in this study was limited, since the conditions for the two first steps, MabSelect SuRe and Capto Q, are more or less generic, while the final Capto adhere step required some evaluation of operating conditions.

Both MabSelect SuRe columns were run five times each to investigate the effects of repeated runs on column performance, as well as to gather enough material for subsequent chromatography steps. The chromatograms obtained were similar. The uniform performance was confirmed by the analytical results. Yields were stable during all cycles, and the contaminant levels were comparable. The HCP level was efficiently reduced, and the ligand leakage low, which is characteristic for MabSelect SuRe. The higher level of ligand leakage in the first cycle, which was detected on both column types, is typical for protein A-based chromatography media. As a result of differences in scale, chromatography systems, and UV detectors, the relative eluate volumes measured in CV were slightly different in the XK and ReadyToProcess runs. Each of the five eluates from the XK runs had volumes corresponding to 1.7 CV, while the five eluates from the ReadyToProcess runs had volumes corresponding to 2.0 CV. Therefore, the sample volumes in the subsequent Capto Q and Capto adhere steps were smaller for the XK runs compared to the ReadyToProcess runs. This difference becomes apparent when comparing the XK and the ReadyToProcess chromatograms from the Capto Q and Capto adhere runs.

The mAb-containing flowthrough and part of the wash were collected. Again, the comparable performance of the XK and ReadyToProcess columns was confirmed by the analytical results. The Capto Q step was characterized by high yield, reduction of HCP, and some reduction of leached ligand and dimer/aggregates. Capto adhere step The mAb-containing flow through and all of the wash were collected. Again, the comparable performance of the XK and ReadyToProcess columns was confirmed by the analytical results. The Capto adhere step had a high yield and efficiently reduced the amount of dimers and aggregates in this study. With this particular mAb, it was necessary to run the column at low pH (pH 5.0) and high salt (0.4 M NaCl) conditions. At these conditions, the HCP removal is limited. Typically, when the mAb allows running at higher pH and lower salt conditions, Capto adhere also removes HCPs.

The performance of ReadyToProcess columns is comparable with established column formats as has been demonstrated in a three-step mAb purification process run in parallel at two different scales: small-scale XK columns and large-scale, prepacked ReadyToProcess columns. The ReadyToProcess columns behave similarly to the XK columns in all aspects studied, demonstrating that the purification process is directly scalable between XK and ReadyToProcess. Multiple cycles (five) have been performed on RTP MabSelect SuRe 2.5 without any detectable changes in column performance.

Smaller disposable columns are available from several sources including Bio-Rad (www.bio-rad.com) and Corning (www.corning.com). There is still an unmet need for inexpensive large disposable columns.

A disadvantage in using resin columns is the large footprint required in their use; compared to this, membrane chromatography employs thin, synthetic, porous membranes that are generally multilayered in a small cartridge, significantly reducing the footprint of the operation. Membranes have the same functional chemical groups to corresponding resins, but they do not need packing, checking, cleaning, refilling, or routine maintenance, and fouled or exhausted modules can be replaced with new ones with minimal process downtime. Sartorius offers a large choice in these membranes.

Membrane Chromatography

Purification of proteins from complex mixtures is a key process in pharmaceutical research and production. But chromatography based on particulate matrices involves lengthy procedures and separation times. Sartobind SingleSep® ion exchange capsules are designed to remove contaminants from therapeutic proteins at accelerated flow rates. This is a direct result of negligible mass transfer effects and is made possible by the >3 μm macroporous membrane. The design allows for robust chromatographic separations and drastically reduced validation costs. Sartobind SingleSep capsules are designed to remove charged contaminants from therapeutic proteins at accelerated flow rates by ion exchange membrane chromatography. The high throughput is a direct result of negligible mass transfer effects and is made possible by the >3 μm macroporous membrane with 4 mm (15 layer) bed height.

Sartobind replaces time-consuming tedious chromatographic steps for many protein and virus applications. The rapid purification on Membrane Adsorbers allows the isolation of protein with high yield up to 100 faster than conventional columns at a flow rate of 20 to 40 bed volumes per minute. (See Figure 8.1.)

The micrograph (Figure 8.2) shows some chromatographic gel beads (average particle size 90 μm) on the surface of the Sartobind Membrane Adsorber. Even at 500-fold magnification, pores of beads are invisible, but the membrane displays a wide pore structure of 3–5 μm size.

Conventional beads keep more than 95% of the binding sites inside the particle. In Sartobind membranes, the binding sites are grafted homogenously as an approximately 0.5–1 μm film on the inner walls of the reinforced and

FIGURE 8.1
Schematic view of viruses binding to functional groups in the membrane pores.

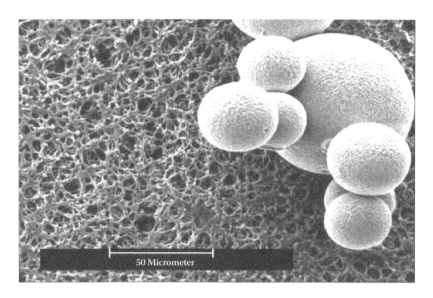

FIGURE 8.2
Sartobind Q membrane and a standard chromatographic matrix.

cross-linked cellulose network. Diffusion time in adsorbers is negligible because of the large pores and immediate binding of the target substance to the ligands. There is no pore diffusion as given in conventional beads but film diffusion anywhere on the microporous membrane structure (Figure 8.3). At convective flow conditions, the movement of the molecules of the mobile phase is directed by pump pressure only. That is why membrane adsorbers feature extremely short cycle times and exceptionally high flow rate and throughput.

There are several available ligands, including

Conventional bead   **Membrane Adsorber**

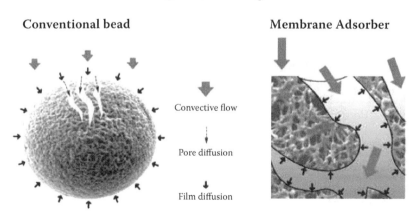

Convective flow

Pore diffusion

Film diffusion

FIGURE 8.3
Existing transport phenomena in conventional beads and membrane adsorbers.

Membrane type	Description	Ligand	Pore size (µm)
Sulfonic acid (S)	Strong acidic cation exchanger	$R\text{-}CH_2\text{-}SO_3^-$	> 3
Quaternary ammonium (Q)	Strong basic anion exchanger	$R\text{-}CH_2\text{-}N^+(CH_3)_3$	> 3
Carboxylic acid(C)	Weak acidic cation exchanger	$R\text{-}COO^-$	> 3
Diethylamine (D)	Weak basic anion exchanger	$R\text{-}CH_2\text{-}N(C_2H_5)_2$	> 3
Phenyl	Hydrophobic interaction (HIC)	Phenyl	> 3
IDA	Metal chelate	Iminodiacetic acid	> 3
Protein A	Affinity	Protein A	0.45
Epoxy-activated	Coupling	Epoxy group	0.45
Aldehyde-activated	Coupling	Aldehyde group	0.45

Source: Sartorius-Stedim

A special advantage in the use of membrane adsorbers is the removal of high-molecular-weight contaminants such as DNA and viruses in monoclonal antibody manufacturing. Such molecules do not readily diffuse into the pores of traditional resins; thus, most polishing steps relying on column chromatography require dramatically oversized columns. These hydrodynamic benefits provide the opportunity to operate the membrane adsorber at much greater flow rates than columns, considerably reducing buffer consumption and shortening the overall process time by up to 100-fold.

Sartorius-Stedim recently introduced single-use, disposable anion-exchange membrane adsorption cartridges that can be used for DNA and host cell protein removal or viral clearance. Likewise, Pall Corporation offers a similar disposable membrane product specifically designed for DNA removal. Other companies including GE Healthcare, Millipore, BioFlash Partners, and Tarpon Biosystems have developed prepacked and presanitized disposable-format chromatography columns. Most of these columns were designed for polishing applications, except Tarpon's, which can be also be used for the capture step in monoclonal antibody (mAb) purification. Capture of target remains a major challenge and contributes to the downstream bottlenecks plaguing many mAb manufacturers.

Virus Removal

Virus contamination is a risk to all biotechnology products derived from cell lines of human or animal origin. Contamination of a product with endogenous viruses from cell banks or adventitious viruses from personnel can have serious clinical implications. According to the European Medicines Agency (EMEA), potential contaminants may have the following characteristics: enveloped or nonenveloped, small or large, deoxyribonucleic acid (DNA) or ribonucleic acid (RNA), and unstable or resistant viruses. Viral safety of licensed biological products must be assured by three complementary approaches:

- Thorough testing of the cell line and all raw materials for viral contaminants
- Assessing the capacity of downstream processing to clear infectious viruses
- Testing the product at appropriate steps for contaminating viruses

A combination of methods—inactivation, adsorption, and size exclusion—is available. The FDA requires demonstration of virus clearance by two methods. Examples of inactivation procedures are solvent and detergent, chemical treatments, low pH, or microwave heating. Methods of adsorption utilize chromatography, and removal by mechanical or molecular size exclusion uses normal (forward) and tangential flow filtration methods. The treatments with solvents/detergent, low pH, or microwave heating all have significant limitations in their ability to inactivate small nonenveloped viruses. Low pH inactivation of murine retroviruses is reported to be highly dependent on time, temperature, and pH, and relatively independent of the recombinant protein type or conductivity conditions outlined. Heating is considered one of the most reliable methods for virus inactivation because of the variation in stability of each virus genome to heat or temperature.

Ion exchange and Protein A chromatography are widely used to remove viruses, and several key studies have been conducted in collaboration with the FDA, yet the responsibility of proving suitability of any method remains the responsibility of the developer.

Viral inactivation such as by ultraviolet exposure is also available, where the virus particles are physically or chemically altered.

Despite the clear demand for downstream processing steps that can provide high levels of viral reduction, few new techniques have surfaced to complement or replace those approaches common in today's biotechnology manufacturing processes. This is particularly true for smaller viruses, such as the parvovirus, which often exhibit resistance to many inactivation strategies such as detergent and heat treatments.

Unfortunately, the risk of failing viral contamination is severe; Table 8.1 shows the action plan in various situations.

Implementing ultraviolet bactericidal (UVC) treatment as one of the orthogonal technologies for virus clearance both for animal cell culture media- and animal cell culture-derived biologicals is recommended. Virus

TABLE 8.1

Action plan for process assessment of viral clearance and virus tests on purified bulk

	Case A	Case B	Case C2	Case D2	Case E2
Status					
Presence of Virus1	—	—	+	+	(+)3
Virus-like particles1	—	—	—	—	(+)3
Retrovirus-like particles1	—	+	—	—	(+)3
Virus identified	Not applicable	+	+	+	-
Virus pathogenic for humans	Not applicable	-4	-4	+	Unknown
Action					
Process characterization of viral clearance using nonspecific *model* viruses	Yes	Yes	Yes		
	Yes	Yes			
Process evaluation of viral clearance using *relevant* or specific *model* viruses	No	Yes	Yes	Yes	Yes
Test for virus in purified bulk	Not applicable	Yes	Yes	Yes	Yes

inactivation with UVivatec® CPV is an integral part of the orthogonal virus clearance technology platform at Sartorius-Stedim Biotech (www.sartorius-stedim.com). This orthogonal technology platform features virus filtration, virus inactivation, and virus adsorption. UVivatec shows efficient (>4 log) inactivation of both small nonenveloped viruses (20 nm) and large enveloped viruses (>50 nm) form a biopharmaceutical feed stream by UV-C irradiation (254 nm) while obtaining product integrity.

UVivatec, a newly developed virus inactivation technology, targets only small and nonenveloped viruses but offers a robust method for inactivating generous and enveloped viruses. This method uses low-dose radiation of UVC at 254 nm, more likely 260 nm (which is more specific to nucleic acids), to destroy the viral nucleic acid while maintaining the structural and functional integrity of the protein of interest. The efficiency of viral inactivation and product recovery is sensitive to the viscosity and absorption coefficient of the protein solution at 254 or 260 nm and its residence time in the radiation chamber. The key to a successful UVC intervention is to introduce it at an early stage of downstream processing so that most aggregates or variant species formed as a result of UVC exposure would be cleared in the subsequent polishing steps.

Planova filters (www.planovafilters) are the world's first filters designed specifically for virus removal. The first Planova filter was launched in 1989 by the Asahi Kasei Corporation, one of the world's leading filter manufacturers. Planova filters significantly enhance virus safety in biotherapeutic drug products such as biopharmaceuticals and plasma derivatives. They exhibit unparalleled performance in clearing viruses, ranging from human immunodeficiency virus (HIV) to parvovirus B19, while providing maximum protein recovery. Planova filters contain a bundle of straw-like hollow fibers. When a protein solution with possible viral contamination is introduced into these hollow fibers, the solution penetrates the fiber wall and works its way to the outside of the fiber.

Within these walls, there is an intricate, three-dimensional network of interconnected void and capillary pores just nanometers in size. Viruses are thus filtered gradually and effectively, while proteins migrate outward with minimal adsorption and inactivation. Planova virus removal works on the principle of size exclusion, unaffected by physical or chemical effects such as adsorption. Therefore, even unknown viruses can be excluded as long as the virus size meets the exclusion specification of the filter. Planova 15N, 20N, and 35N filters are offered in 0.001, 0.01, 0.12, 0.3, 1.0, and 4.0 m^2 sizes. The 4.0 m^2 filter reduces the number of filters needed for a manufacturing cycle and shortens cumulative integrity test time. The PVDF media Planova BioEX filter is offered in 0.001 and 1.0 m^2 sizes. Additional sizes, including 0.0003, 0.01, 0.1, and 4.0 m^2, are under development to extend the Planova BioEX line.

ChromaSorb membrane adsorber, an innovative, single-use, flow-through anion exchanger from Millipore (www.millipore.com) , is designed to remove trace impurities, including host cell protein (HCP), DNA, endotoxins, and viruses. This device provides the greatest levels of impurity binding at the highest salt concentrations for mAbs and other protein purification steps. The ability to reliably perform at greater salt concentrations significantly reduces buffer volumes. Compared to traditional column-based anion exchange resins, ChromaSorb membrane adsorber reduces the validation requirements, capital equipment expenditures, time, and labor. The ChromaSorb membrane consists of a uniform 0.65 μm pore structure, ultrahigh-molecular-weight polyethylene (UPE) membrane coated with a cationic primary amine ligand for high binding strength for negatively charged impurities.

Mobius FlexReady Solution for Virus Filtration from Millipore (www.millipore.com) provides the best combination of single-use assemblies (Flexware assemblies), innovative separation devices, and process-ready hardware specifically designed for virus filtration. It is preassembled and pretested, and integrates easily into any process. Designed for applications where feed volumes range between 1 and 200 L at protein concentrations from 5 to 10 g/L, the Mobius FlexReady Solution for Virus Filtration features Millipore's next-generation virus clearance solution, Viresolve Pro Solution.

Designed for applications where feed volumes range between 1 and 200 L at protein concentrations from 5 to 10 g/L, the Mobius FlexReady Solution for Virus Filtration features Millipore's next-generation virus clearance solution, Viresolve Pro Solution. The benefits include a unique low pulsation peristaltic pump, constant pressure operation using pressure feedback and pump speed control, process end-point via percent flux decay, cumulative volume/weight or processing time, innovative low dead volume t-connectors, enabling the use of traditional pressure transducers, presized sterilizing-grade filter and product collection assembly, ease of operation with an intuitive touchscreen interface, and user-defined process and alarm set points.

Virus filtration with Virosart® CPV is an integral part of the orthogonal virus clearance technology platform at Sartorius Stedim Biotech (www.sartorius-stedim.com). This orthogonal technology platform features virus filtration, virus inactivation, and virus adsorption. Virosart® CPV shows efficient removal of both small nonenveloped viruses (20 nm) and large enveloped viruses (>50 nm) from a biopharmaceutical feed stream by size exclusion. The double-layer 20 nm PESU membrane of Virosart® CPV features excellent flow rates and superior capacity. The filter offers highest viral safety over the entire flow decay profile up to 90%. Virosart® CPV filters are being tested for integrity using a water-based integrity test. All filters have been validated for 4 log10 removal of small nonenveloped viruses using bacteriophage PP7 as a model virus. Each individual Virosart® CPV filter is autoclaved and integrity-tested during manufacturing assuring highest product reliability. Virosart® CPV provides highest viral safety to the biopharmaceutical product. This filter retains more than 4 log10 of small nonenveloped viruses (porcine parovirus [PPV] or mouse minute virus [MVM]) and more than 6 log10 of large enveloped viruses (e.g., MuLV). Based on the unique double-layer 20 nm polyethylene sulfone (PESU) membrane, Virosart® CPV provides excellent flow rates and superior capacity. This filter offers highest viral safety over the entire flow decay profile of up to 90%.

Scale-down work is being realized using the Virosart® CPV Minisart (5 cm² capsule) to enable filtration work for flow and capacity studies as well as for GLP virus spiking studies. Scale-up studies are being performed using this capsule's size 9 with filter area of 2.000 cm² to reliably scale up into larger-scale manufacturing. Typical batch sizes of products being subject to nanofiltration with this Virosart® CPV capsule size 9 are 5 to 50 L.

Pall (www.pall.com) offers Kleenpak Nova filters with either in-line or T-style configurations. The T-style configuration is ideal for manipulating multiple filters in series or in parallel configuration. Kleenpak Nova Capsule filters incorporate either a 10 in. (254 mm), 20 in. (508 mm), or 30 in. (762 mm) length standard Pall cartridge filter, which have traditionally been installed in stainless steel housings. In applications where a particular filter is already specified, the user can switch from a stainless steel housing to a fully disposable assembly with minimal requalification. This means the extensive range of sterilizing-grade and virus filters currently available from

Pall can easily be provided as a capsule filter, including low binding, high-flow Fluorodyne® II polyvinylidene difluoride (PVDF) filters, Ultipor® N66 and positively-charged Posidyne® nylon 66 filters, Supor® polyethersulfone filters, and Ultipor VF DV20 and DV50 virus removal filters. Kleenpak Nova capsules are especially suited to pilot- and process-scale applications. They can be autoclaved or sterilized by gamma irradiation and can be supplied as part of presterilized processing systems such as a filter/tubing/bag set. Additionally, the disposable Kleenpak sterile connector allows for the dry connection of two separate fluid pathways, while maintaining the sterile integrity of both. The connector consists of a male and a female connector, each covered by a vented peel-away strip that protects the port and maintains the sterility of the sterile fluid pathway. Different connector options are available to allow for the connection of 15.8 mm (5/8 in.), 13 mm (½ in.), 9.6 mm (3/8 in.), or 6 mm (¼ in.) nominal tubing. Kleenpak sterile connectors can either be autoclaved up to 130°C or gamma-sterilized.

Viral clearance studies require viral spiking and many controls and validation protocols that might be difficult to conduct in-house. Several contract laboratories are available to perform these studies, and generally the author recommends outsourcing this phase of development:

AppTec Laboratory Services, www.apptecls.com

Bioreliance, Inc., www.bioreliance.com

Charles River Laboratories, www.criver.com/products/biopharm/biosafety.html

Inveresk Research Group, Inc., www.inveresk.com

Microbix Biosystems, Inc., www.microbix.com

Q-One Biotech, Ltd., www.q-one.com

Texcell – Institut Pasteur,
www.pasteur.fr/applications/dri/French/Texcell.html

Buffers

Buffer preparation and storage requires large volume handling, generally about 10 times the upstream media volume; some resins such as Protein A require large buffer volumes. While there are several standard buffers, specific pH and electrolyte requirements make standardization of buffers difficult, and these are often process specific. While several development projects underway promise to reduce the volume of buffer needed, the current systems are likely to stay the same for some time. The main efforts in this regard include reducing the process steps.

The reduction in buffer volume is necessary because of the high cost of the water used in preparing buffers. While it is acceptable to use Purified Water USP to prepare buffers, many large manufacturing operations used water for injection (WFI) instead because often this is the only choice available. Recently, a new manufacturing facility in Chicago (www.theraproteins. com) qualified a double reverse osmosis the electro-deionization (EDI) water system for biological manufacturing, saving almost 90% of the cost of water used, from about US$3/L for WFI based on the stainless steel system to less than 5 cents per liter.

Buffer storage was one of the first unit operations to transition to single-use systems. Recent analyses have confirmed that there is a clear economic advantage to this methodology over traditional hard-piped systems. Buffer mixing, however, continues to rely on more traditional technology. This is partially due to the scale of many buffer preparation processes and partially due to a reliance on existing infrastructure. However, as new facilities are commissioned and as new technologies are introduced that limit the volume of buffers required, single-use mixers are being chosen over traditional technologies. The shift to single use is driven by the needs to minimize capital investment, enable more rapid process setup, reduce downtime, and provide increased flexibility.

The largest disposable holding bags currently have a 3,000 L capacity (Sartorius Palletank), while mixing with disposables is limited to 5,000 L (Hynetics Disposable Mixing System), with more systems available at the 1,000 to 2,000 L scale. Preconfigured, disposable stand-alone systems for buffer preparation have been launched recently with 500 and 1,000 L capacities (e.g., Mobius). However, most of these packaged systems are very expensive and do not add any real value to the bioprocess since a simple system such as a Dixie Poly Drum 330 Gallon Economy Tank with a polyethylene (PE) liner would do the job as well (the cost of the entire system is less than US$1,500); Class VI USP liners are readily available even in a sterile state, though that is not necessary. For companies starting out, it is recommended that they develop their own configured system for storage of buffers and also use the same tanks for mixing purposes. Often, a built-in system of a cage that will accommodate a liner is sufficient for the purpose.

As an example of integrated systems, for buffer preparation steps, GE Healthcare (www.gelifesciences.com) offers its WAVE™ Mixer. The WAVE Mixer provides efficient mixing in a sterile, sealed bag by an innovative method. Instead of using a pump or invasive impeller to induce circulation flow, it uses waves generated in the liquid by a precisely regulated rocking motion. The system has been optimized for extremely efficient mixing and dispersion. Wave motion moves large volumes of fluid and disperses solids. The WAVE Mixer eliminates the need for a mixing tank and conventional mixer. This also eliminates equipment cleaning, sterilization, and validation.

The WAVE Mixer system comprises two main components: a special rocking platform that induces a wave motion in the liquid without an impeller or other invasive mixer and the M*Bag™, which contains the ingredients to be mixed and dissolved. The unique M*Bag is made of a multilayer laminated clear plastic designed to provide high mechanical strength. A large screw cap port allows powders or other solids to be easily poured into the bag. Probes to measure pH and conductivity can be inserted. A large outlet port allows the M*Bag to be drained completely.

Standard systems are available for 20 and 50 L bags. These can be used to mix volumes from 1 to 35 L of liquid. Larger systems up to 500 L liquid volume are available. The WAVE Mixer principle has also been used for the mixing of materials in custom-shaped rigid containers.

Biological and particle contaminants present in buffers can have a large impact on process efficiencies and final product quality. Therefore, normal flow filtration is one of the first steps (after dissolution) in any buffer preparation process. Buffer filtration is key to protection of chromatography columns and ultrafiltration operations and to the production of an endotoxin-free final product. Buffer filtration also aligns with FDA guidelines on sterile drug products that advise to reduce and control the bioburden across the process.

Buffer filters should be chosen based on the following characteristics:

- Validated retention of bacteria
- Broad chemical compatibility
- High permeability
- Physical robustness

For filtration of sterile and reduced-bioburden buffers, GE Healthcare offers ULTA™ Pure SG (for buffer sterilization) and ULTA Prime CG (for bioburden reduction). The ULTA family of filters uses membranes with industry-leading permeability that consistently outperforms competitive offerings. Both filter grades are constructed using a polyethersulfone 0.2 micron membrane, which is physically robust and chemically resilient, so they perform reliably regardless of the buffer being prepared. Additionally, both filters employ a final membrane that is validated for bacterial retention using the ASTM F838-05 methodology (LRV > 7 for ULTA Pure SG and LRV > 5 for ULTA Prime CG). ULTA filters are available in a wide variety of cartridge and capsule formats with surface areas ranging from 450 cm^2 to 1.5 m^2 in a single device.

Fluid Management

Once prepared, buffers are transferred to downstream processing stations when using larger volumes, on carts; or else through a tubing system with the help of peristaltic pumps. In Chapter 6, a review of the connectors, both aseptic and otherwise, was presented.

ReadyCircuit assemblies comprise bags, tubing, and connectors. Together with ReadyToProcess filters and sensors, ReadyCircuit assemblies form self-contained bioprocessing modules that maintain an aseptic path and provide convenience by removing time-consuming process steps associated with conventional systems. Bags, tube sets, filters, and related equipment can be secured in appropriate orientations for efficient operation using the ReadyKart mobile processing station. With an array of features and optional accessories, the ReadyKart is designed to support a variety of process-specific, fluid-handling needs.

ReadyToProcess Konfigurator lets one design fluid-handling circuits with ease online. One enters the parameters to generate the design one needs; and includes fast output of piping and instrument drawings (P&ID) drawings and convenient Bill of Materials for simplified ordering.

ReadyMate connectors are genderless aseptic connectors that allow simple connection of components, maintaining secure workflows and sterile integrity. Additional accessories such as a tube fuser and sealer of thermoplastic tubing support secure aseptic connectivity throughout the manufacturing process.

ReadyToProcess filters are a range of preconditioned and ready-to-use cartridges and capsules for both cross-flow and normal flow filtration operations. It is factory prepared to WFI quality for endotoxins, total organic carbon (TOC) and conductivity and sterilized via gamma radiation. They enable simpler and faster bioprocessing with maximum safety.

Bioseparation

Once the expression phase ends, the first step of bioseparation starts.

In the case of bacterial expression, this would involve a continuous flow centrifuge (e.g., New Brunswick Classroom Engineering Process Assistant (CEPA) centrifuge, www.nbsc.com/cepa.aspx). Smaller volumes can be processed in other standard centrifuges taking about 4–6 L in each run. At this point in the development, we do not have a disposable centrifuge option except the Centritech Cell II by Pneumatic Scale Angelus (www.pneumatic-scale.com/) that can process up to 120 L/h. It will require running several centrifuges in parallel to process a typical 2,000–4,000 L run; this centrifuge, however, is capable of separating animal cells as well.

The next stage for handling the cell mass would be to use an enzyme method, a bead method, sonication, detergent, or solvent method. Homogenization using a French Press is most common for large-scale processing. At this time, there are no disposable mechanical systems available.

And perhaps there is no need for it if the process can be isolated and, more particularly, if a single product is handled in a single facility.

Depth Filtration

Cell removal by filtration leaves media and its entire component in the same volume, and this requires depth filtration. Depth filters are filters that use a porous filtration medium to retain particles throughout the medium, rather that just on the surface of the medium. These filters are commonly used when the fluid to be filtered contains a high load of particles because, compared to other types of filters, they can retain a large mass of particles before becoming clogged. These filters are discussed in Chapter 10.

The performance of depth filters is largely dependent on the colloid content of the bioreactor offload and the cell debris removal capacity of the upstream centrifuge. Usually, depth filters are operated with a constant flow of 100–200 $L/(m^2 \cdot h)$ and up to 150 Lfeed/m^2 of filter depending on the composition of the feed stream. The Millipore Millistak+ Pod system has a maximum capacity of 33 m^2 filter area, resulting in a batch capacity of 3000–5000 L. The Millipore Mobius FlexReady process equipment supports offers a larger 55 m^2 filter area. Since the washing of these filters requires very large volumes of buffers, appropriate size holding tanks are required that can be lined with disposable PE liners.

Ultrafiltration

Ultrafiltration (UF) is a variety of membrane filtration in which hydrostatic pressure forces a liquid against a semipermeable membrane. Suspended solids and solutes of high molecular weight are retained, while water and low-molecular-weight solutes pass through the membrane. This separation process is used in industry and research for purifying and concentrating macromolecular (10^3–10^6 Da) solutions, especially protein solutions. Ultrafiltration is not fundamentally different from microfiltration, nanofiltration, or gas separation, except in terms of the size of the molecules it retains. Ultrafiltration is applied in cross-flow or dead-end mode, and separation in ultrafiltration undergoes concentration polarization.

Diafiltration is a membrane-based separation that is used to reduce, remove, or exchange salts and other small-molecule contaminant from a process liquid or dispersion. In batch diafiltration, the process fluid is typically diluted by a factor of two using "clean" liquid, brought back to the original concentration by filtration, and the whole process is repeated several times to achieve the required concentration contaminant. In continuous diafiltration, the "clean" liquid is added at the same rate as the permeate flow.

Cross-flow filtration (also known as tangential flow filtration) is a type of filtration (a particular unit operation). Cross-flow filtration is different from dead-end filtration in which the feed is passed through a membrane or bed, the solids being trapped in the filter and the filtrate being released at the other end. Cross-flow filtration gets its name because the majority of the feed flow travels tangentially across the surface of the filter, rather than into the filter. The principal advantage of this is that the filter cake (which can blind the filter) is substantially washed away during the filtration process, increasing the length of time that a filter unit can be operational. It can be a continuous process, unlike batch-wise dead-end filtration. This type of filtration is typically selected for feeds containing a high proportion of small-particle-size solids (where the permeate is of most value) because solid material can quickly block (blind) the filter surface with dead-end filtration. Industrial examples of this include the extraction of soluble antibiotics from fermentation liquor.

Ultrafiltration and diafiltration steps are used to concentrate and change the buffer of a solution. During final formulation, ultrafiltration/diafiltration is used to transfer the active pharmaceutical ingredient to a stabilizing environment and to achieve the correct concentration. Up to 300–5,000 L may need to be processed, depending on whether the column eluates can be fractionated. Membranes with a 30 kDa molecular weight cutoff are often used to retain antibodies, and the process intermediate is concentrated and washed with 5× volumes. Modules of up to 3 m² are available that can process 200 L/ (h m²). Several disposable systems are available (Scilog, Millipore) for a limited filter area (up to 2.5 m²), but larger systems that might replace existing reusable systems with 14 m². Because there are already disposable modules and pumps available, its logical to carry the filtration steps in a closed system.

Integrated Systems

ÄKTA ready (www.gehealthcare.com) is a liquid chromatography system built for process scale-up and production for early clinical phases. The system operates with ready-to-use, disposable flow paths and as a consequence, cleaning between products/batches and validation of cleaning procedures is not required. ÄKTA™ready is a liquid chromatography system built for process scale-up and production for early clinical phases. System meets Good Laboratory Practices (GLP) and current Good Manufacturing Practices (cGMP) requirements for Phase I–III in drug development and full-scale production and provides improved economy and productivity due to simpler procedures. Single use eliminates risk of cross-contamination between products/batches, facilitates easy connection to and operation with pre-packed ReadyToProcess™ columns and other process columns, and enables scalable processes using UNICORN™ software.

Studies conducted using the foregoing ReadyToProcess system by GE to manufacture a monoclonal antibody and comparing it with the traditional system shows a reduction in the total process time by half from almost 50 hours to 24 hours. Most of the time savings comes from obviating interim validation of equipment. The purity levels of the product obtained are generally comparable. However, in this particular example, the cost of materials is substantially higher when using the ReadyToProcess system; how much of it gets reduced by the time savings would depend on the cost of time to the individual processor.

Step	ReadyToProcess	Traditional	% Time saving
Upstream	Wave Bioreactor	Stirred tank	0
Capture	ReadyToProcess Protein A	AxiChrom 70 column with Protein A	55
Buffer Exchange, UF	ReadyToProcess hollow fiber cartridges	Kvick cassettes	55
Polishing	ReadyToProcess prepacked column, Capto	AxiCrom Capto	66
Formulation	ReadyToProcess hollow fiber cartridges	Kvick cassettes	30

While the foregoing example provides a very good view of the state of the art today, there remain many hurdles to providing a true cost-effective system. One way of achieving this goal would be to combine the upstream and downstream processes, as suggested elsewhere in this book.

The foregoing system, however, can be further modified by using Scilog's single-use ultrafiltration system. The SciPure 200 (www.scilog.com) is a single-use bioprocessing platform for the automation, optimization, and documentation of tangential flow filter (TFF) Applications. The system has been designed to meet cGMP and 21 CFR Part 11 standards for data collection and security as a stand-alone device with the ability to create and execute discrete or batch operations for filling, concentration, diafiltration, and normalized water permeability (NWP). User-selectable end points and alarms enable hands-free operation and ensure safe, consistent process performance. The patented proprietary technology enables the SciPure 200 system to respond to sensor feedback and thus maintain the user-defined flows and transmembrane pressure (TMP) simultaneously. All wetted flow-path manifold components of the SciPure 200 are considered single-use consumables.

SciLog (www.scilog.com) recently developed a fully automated, single-use purification platform that purportedly improves downstream processing efficiencies and may help cut costs. Other companies are developing disposable-format expanded bed adsorption and high-capacity monolith and membrane adsorbers to improve capacity during the capture.

MayaBio (www.mayabio.com) has recently developed a separative bioreactor where the binding resin is added directly to the biroeactor contained in a filter pouch with 30-micron-size screen; once the protein in the media is absorbed onto resin or membrane, the bioreator is drained out. The drain is closed, the resin/membrane is washed with buffers to remove debris, and is finally washed with an eluting buffer to a volume that is generally less than 5% of the original media volume. This process eliminates several steps: bioseparation, ultrafiltration, and buffer exchange. If the resin/membrane used to capture protein can be used in the column, then this approach eliminates the lengthy procedure of column loading as well.

9

Filling and Finishing Systems

I never see what has been done; I only see what remains to be done.

Buddha

There are two stages where a biological product requires filling in containers for long-term storage. First, it comes at the end of the downstream cycle when a purified solution of a protein drug is ready to be filled into containers as a bulk drug that will be shipped out to companies to formulate and finish the product in a dosage form for the end customer. The second stage of filling is to formulate the bulk and fill into end-customer dosage form, a syringe, vial, or ampoule.

The bulk product is generally labeled as a pharmacopoeia product, such as "Erythropoietin Concentrated Solution EP," and would thus comply with all requirements of labeling as required in the pharmacopoeia. There is no need for any formulation additives, and the last buffer exchange in the downstream processing likely already brought it into a formulation that is stable. This stage of filling is conducted as a continuation of downstream processing. Sterile serum bottles are available for packaging the product (e.g., from Thomas Scientific, www.thomassci.com) in sizes from 25 mL to 9 L. These solutions can alternately be filled in sterile flexible bags. It would be prudent to fill them using a sterile filter as the final stage, perhaps a virus-clearing filter.

The second stage of filling protein solutions is more demanding. The final production step is transferring the new medium from the transfer vessel or bags and into vials for distribution. Traditionally, the final fill operation consisted of stainless steel equipment connected via reusable valves, rigid tubing, and steel pipes. Again, this equipment requires validation and must be subjected to a cleaning-in-place (CIP) cycle after each filling cycle is completed. Today, many process engineers are designing this operation with single-use tubing assemblies in place of stainless steel piping to reduce sterilization time and cost.

One example of integrating single-use systems in a final fill operation is for simplifying mobile transfer tanks. These tanks with disposable liners are designed to transfer a product from formulation suites to storage areas and ultimately to filling suites. To allow sterile connection to and from these vessels, designers traditionally add three-way valve assemblies to fill and drain ports to facilitate sterilization-in-place (SIP) operations. The design of these three-way valves makes it difficult to validate cleaning procedures.

Replacing these heavy three-way valve assemblies with single-use tube sets and connectors eliminates cleaning, validation, and maintenance.

Single-use tubing assemblies can either be attached prior to equipment sterilization with single-use SIP connectors (used as either steam access or condensate drainage sites), or steamed separately, just prior to fluid transfer. For vessel outlet, combining a number of single-use components into the transfer line can create a robust system to ensure product safety. For example, outlet transfer lines could incorporate a single-use SIP connector to attach to the sterile holding tank. Then, a through-the-wall fluid transfer system is used to bring a portion of the transfer line into the filling suite. Next, a sterile connector is used to attach the transfer line to a separate portion of the transfer line that has already been steamed onto the filling machine with a single-use SIP connector. Finally, disconnecting the transfer lines using a quick disconnect coupling that has been validated as an aseptic disconnect enables the processor to confidently make an aseptic disconnection from the storage vessel or bag.

The leader in sterile product filling systems remains Bosch Packaging, among others. The following information was developed by Robert Bosch Packaging Technology, Inc. (www.boschpackaging.com).

Robert Bosch Packaging Systems

Use of disposable components in product downstream processing and final fill operations is increasing as the technology for performing these steps in a single-use mode also increases. There is a high demand for systems that support single-use purification, formulation, and filling operations. There are several drivers for this. First is the desire to realize increased processing efficiency through the elimination of preparative steps such as CIP and SIP for product-contact equipment and parts. For example, presterilized single-use tubing and bags can be used to replace stainless steel piping and tanks that have to be cleaned and steamed between uses. Second is the reduction of validation efforts related to the product path, in particular, the elimination of cleaning validation. Products that are hard to clean, or are highly potent or toxic, often require dedicated product-contact parts. This is because existing cleaning processes are inconsistent or simply do not work to move certain products to safe levels. Third is containment of toxic products. Disposable systems can be removed, bagged, and disposed of without breaking connections and exposing the environment to the product. Fourth is the desire to match existing single-use upstream processes that are available for new products, particularly, biopharmaceuticals and other protein-based drugs.

In the past, filling line equipment was commonly dedicated entirely to a single product. However, this approach is no longer economically feasible except for the highest-volume drugs that support nearly continuous filling operations. Most filling lines today support multiproduct operations. Traditional multiproduct operations require validation of the level of product carryover after cleaning operations to ensure subsequent products are not contaminated. Certain product-contact parts are hard to clean to acceptable levels, and are therefore dedicated to specific products.

An alternative to filling equipment dedication is the use of single-use parts and assemblies. These systems can include components such as bulk product bags, capsule filters, silicone tubing, and other plastic fittings and parts. Many of these parts can already be purchased precleaned and presterilized, and double or triple bagged for easy use within clean rooms. However, filling operations are critical enough to require entire single-use systems be assembled and sterilized together rather than having to piece individual components together at the point of use. The dosing system used for filling also affects the characteristics of the single-use components.

The presterilized, single-use concept has already been realized with several off-the-shelf filling systems using peristaltic or gravimetric dosing. Many existing systems, however, are designed for low-speed, small-batch filling operations. Scale-up of these systems for high-speed filling have created technical obstacles that include a relatively slow dosing speed, lower filling accuracy and precision, and difficulty dosing products with variable temperature and viscosity characteristics.

Peristaltic dosing is ideally suited for disposables, as peristaltic tubing is used for much of the product path. Single-use peristaltic systems typically comprise a product hold bag, supply tubing, and a filling needle, which are bagged together and sterilized using gamma irradiation. The assembly is removed from the bag and connected to the filling system, which can be as simple as a single peristaltic pump, immediately before use. High-quality peristaltic pumps can be very precise at dosing water-like solutions. However, the tubing directly influences accuracy. The tubing that is located in the pump head changes shape over time due to wear, so accuracy drift is common. Characterizing and compensating for the drift is required. Some peristaltic pumps also dose at slower speeds, which means that high-speed peristaltic systems require more pump heads to dose at the same rate as the equivalent piston, time pressure, or rolling diaphragm systems.

Gravimetric dosing uses optical sensors to dose a given volume based on a calculation of the interior volume of a given length of tubing or glass. The entire product path is supplied as a single-use, presterilized assembly. Accuracy and precision of the system with water-like solutions is comparable to other dosing systems. However, dosing speed and accuracy can be directly related to fluid temperature and viscosity. Dosing time is based on the speed at which a liquid will flow through the tubing based on gravity. Thicker solutions flow more slowly and are therefore dosed at a slower rate. Relatively small temperature changes over the course of a filling event can affect product viscosity enough to have a significant effect on the volume filled. Similar to peristaltic systems, more pump heads are required to dose at the same rate as the equivalent piston, time pressure, or rolling diaphragm systems.

The current single-use, presterilized dosing systems are based on scaling-up technologies designed and used for small-scale filling operations. However, a better approach is to convert existing high-speed dosing technology to single-use. The three most common commercial systems are piston pumps, rolling diaphragm pumps, and time pressure dosing. All three of these systems require significant technical improvements and modifications to be converted to disposable use.

Piston pumps rely on a precise physical tolerance between the pump body and piston to provide dosing accuracy and to ensure the product does not leak during use. Pump bodies and pistons are commonly matched when they are fabricated to ensure they do not gall during use. Existing piston pumps for pharmaceutical dosing can be made using stainless steel or ceramic components. Neither material can be used to make a disposable pump due to the high cost of manufacture. Plastic components are an alternative but cannot be fabricated to the correct tolerances to ensure accuracy. Excessive wear and leaking would also be issues. A catastrophic loss of function will likely result without the use of o-rings, a lubricant, or both to separate the moving plastic surfaces. Plastic particles, elastomeric particles in the case of o-ring use, or lubricant will also be shed by the pump and introduced into the product stream (see Figure 9.1).

The rolling diaphragm pump, originally developed by TL Systems (now part of Bosch), comprises a stainless steel pump with a diaphragm. A headpiece and diaphragm make up the liquid chamber. Dosing occurs by actuating a piston that is attached to the diaphragm. It is very similar to piston dosing, only the diaphragm keeps the product from contact with the piston and other internal components. The only stainless steel part in contact with the product is the headpiece (see Figure 9.2). Unlike the piston pump, however, these surfaces are separated from the fluid path by the diaphragm, so contamination of the product stream will not be an issue. The tolerances for each part are not as critical as with piston pumps, as dose accuracy is related to accurate piston stroke while at the same time maintaining consistent dimensions in the fluid chamber.

Time pressure systems are designed to dispense using a pressurized product supply and timed valve openings (see Figure 9.3). A portion of the product path from the product supply manifold to the filling nozzles is made of elastomeric tubing. This tubing is used in association with an automatic tubing pinch mechanism to create the valve. The use of disposable tubing seems to make the system a good candidate for a single-use system. However, this is not the case. These systems often use a small surge tank for the product supply, and this tank must be pressurized up to 10 psig or more for the system to function. Replacing this tank with a bag would require that the bag be pressurized beyond its normal design pressure. There is currently no good

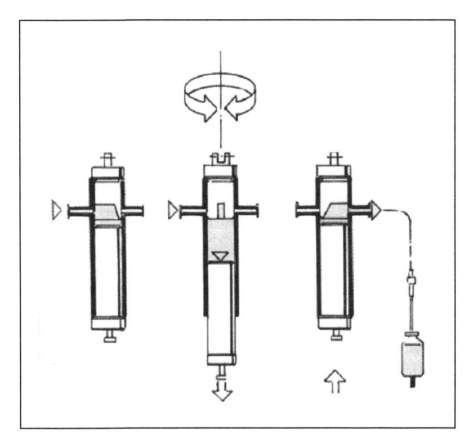

FIGURE 9.1
Piston pump cross section as used in Bosch machines.

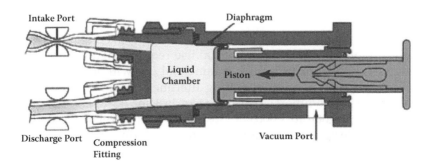

FIGURE 9.2
Rolling diaphragm pump cross section as used in Bosch machines.

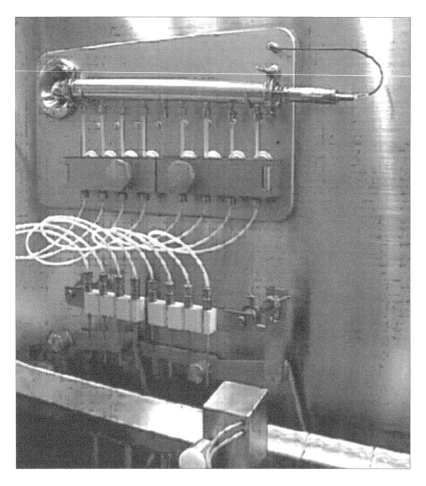

FIGURE 9.3
Time pressure system of Bosch machines.

solution for pressurization of a surge bag system for use with time pressure filling.

Ensuring a single-use system dispenses at high speed and at the same time is durable enough for commercial use requires rigorous testing. No dosing system is appropriate for commercial use without proof of accuracy and precision over its operating lifetime. The maximum intended run duration for commercial systems can last for up to a week or more and involves 500,000 to 1,000,000 dosing cycles per station. This is well beyond the design specification of many existing single-use dosing systems.

A limitation to many current single-use presterilized dosing systems is the plastic filling needle. The current plastic needles are not designed for commercial filling operations. Most are too wide to penetrate small containers and/or are too short to perform bottom-up filling. Bottom-up filling, where the filling needle penetrates the container and is drawn out during dosing, is common with high-speed filling to reduce product splash and foaming. The plastic needles are also not shaped to fit correctly within needle holders on common commercial filling systems. Custom fixtures are required to use them on existing machines.

High-speed filling requires needles made to very tight tolerances, particularly the needle diameter, as this has an influence on dosing accuracy and precision. Because high-speed needles travel during and after dispensing, needle drip between doses has to be eliminated. Precise needle opening size and opening shape is also required. Substituting plastic needles with ones made from stainless steel can solve most of these issues, but is too expensive for single-use assemblies.

The PreVAS family of aseptic single-use dosing systems by Bosch represents a major step in dosing system technology. PreVAS is the first completely preassembled and presterilized dosing system available for the clinical and production pharmaceutical and biotech filling market that is supplied with supporting validation documentation. This allows a risk-free scale-up of filling operations in a single-use format. PreVAS removes several risk factors from the filling operation. There are no complicated cleaning procedures and validation protocols required, and the entire system is quickly installed and made operational (Figure 9.4).

PDC Aseptic Filling Systems

The PDC Aseptic Filling Systems (www.lpsinc.net/pdc) specializes in the design and manufacture of innovative filling solutions used by major biopharmaceutical companies (Figure 9.5). Through years of development, testing, and validation by major biopharmaceutical companies, PDC's technology allows for aseptic filling within any cleanroom classification, at a fraction of fixed stainless steel system cost and manifold filling cost. Further, through the use of proprietary disposable fill system liners, customers have a fully disposable option eliminating the risk of cross-contamination and eliminating much of the cost of cleaning and documentation at all levels of pharmaceutical production.

PDC manufactures aseptic filling systems featuring a unique disposable sterile fill line. Aseptic filling systems are available to fill bottles, bags, or drums from 5 mL up to 1,000 L. PDC's disposable filling lines offer single-use noncontact aseptic filling that can be custom-designed for a variety of biopharmaceutical filling applications.

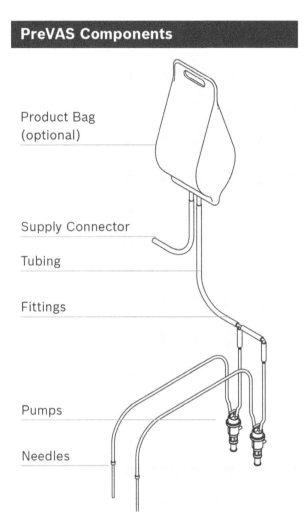

PreVAS Components

Product Bag
(optional)

Supply Connector

Tubing

Fittings

Pumps

Needles

FIGURE 9.4
The PreVas system.

Other solutions for filling the bulk are available from Sartorius, Pall, and Millipore.

The Integrity™ LevMixer® system is a mobile, flexible mixing system that allows efficient and reproducible single-use mixing of a wide range of volumes in a broad series of applications ranging from buffer preparation to final formulation. The LevMixer system is engineered for use with ATMI single-use mixing bags in cGMP-certified cleanrooms without complex instrumentation to control the mixing process. The LevMixer consists of an interchangeable superconducting drive unit and proprietary levitating impeller-based disposable mixing bags fitted into containers on either a

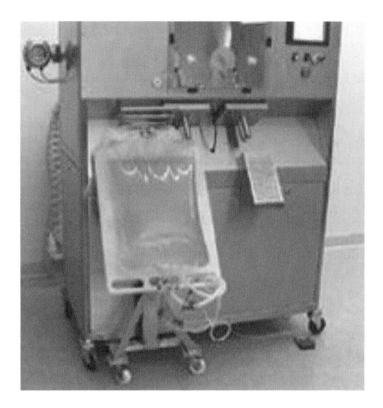

FIGURE 9.5
The PDC filling system.

portable dolly or a floor-mounted tank. Once properly charged and coupled with the mixing bag, the activation of the motor induces levitation and rotation of the in-bag impeller resulting in effective mixing action inside a hermetically sealed bag. Coupling of the in-bag impeller with the drive motor requires no dynamic seals or shaft penetration inside the bag. The drive motor is enclosed on a portable cart that can be easily disconnected from the bag and reconnected to another mixing bag, allowing mixing in multiple bags of various sizes with a single drive unit. As with all ATMI LifeSciences' single-use mixing systems, the LevMixer utilizes disposable mixing bags made from Integrity TK8 bioprocess film. The product-contacting layer of TK8 film is blow-extruded in-house by ATMI under cleanroom conditions using medical-grade ultra-low-density polyethylene resin. It is then laminated to create a gas barrier film of exceptional cleanliness, strength, and clarity that is animal-derived component free (ADCF) and complies fully with USP Class VI requirements.

Palletank® for LevMixer® is a stainless steel cubical container designed to perfectly fit with the Flexel® Bags for LevMixer® with its integrated impeller. It includes a railed port for coupling the mobile LevMixer® drive unit with the Flexel® Bags for LevMixer® and a clamp holder to facilitate powder transfer. The hinged door allows easy installation of the bag system, whereas the front bottom gate facilitates easy tubing installation and access. Windows on lateral and rear sides enable the user to visually control the mixing process. The cubical shape improves the mixing efficiency and offers scalability from 50 to 1,000 L.

Pall's Allegro™ 3D single-use mixers are impellor-based systems for large-volume applications including vaccine and drug formulation, and mixing viscous fluids. Available as 200 L systems for working volumes from 50 to 200 L, Allegro 3D mixers are also ideal for a wide range of liquid–liquid and solid–liquid mixing applications such as compounding, formulation, buffer and media preparation, and pH/conductivity adjustment. Front-loading of the disposable mixer assembly eases setup. Mixing data is available for a wide range of industry standard applications, demonstrating fast and complete mixing even for preparations requiring high concentrations and large additions of solids.

Summary

There are tremendous advantages to the use of single-use, presterilized dosing systems for commercial filling operations. Increased processing efficiency through the elimination of preparative steps such as CIP and SIP, reduction of validation efforts including elimination of cleaning validation, containment of toxic products, and matching existing single-use upstream processes are all compelling arguments for these systems for product filling operations. However, significant technical achievements must be realized before a system can be scaled for high-speed filling operations.

10

Filtration

I wish all teenagers can filter through songs instead of turning to drugs and alcohol.

Taylor Swift

Except for steel meshes in bulk manufacturing of nonsterile dosage forms, filters are rarely reused in the pharmaceutical industry. They take varied forms—from muslin cloth to paper filters to membrane cartridges. Disposable filter devices in biological manufacturing were the earliest changes that went disposable mainly because of the problems with cleaning them; the cost of these parts has always been reasonable.

There are a multitude of filter designs and mechanisms utilized within the biopharmaceutical industry. Prefilters are commonly pleated or wound filter fleeces manufactured from melt-blown random fiber matrices. These filters are used to remove a high contaminant content within the fluid. Prefilters have a large band of retention ratings and can be optimized to all necessary applications. The most common application for prefilters is to protect membrane filters that are tighter and more selective than prefilters. Membrane filters are used to polish or sterilize fluids. These filters need to be integrity testable to assess whether or not they meet the performance criteria. Cross-flow filtration can be utilized with micro or ultrafiltration membranes. The fluid sweeps over the membrane layer and therefore keeps it unblocked. This mode of filtration also allows diafiltration or concentration of fluid streams. Nanofilters are commonly used as viral removal filters. The most common retention rating of these filters is 20 or 50 nm.

Dead-End Filtration

Dead-end filtration operates on the principle of passing a fluid feed stream through a filter device by means of a pressure drop, usually applied by either a pump or compressed gas pressure before the filter device. All contaminants larger in size than the pore size of the filter media are retained by the filter material and will finally cause a filter blockage by plugging its channels or pores. The dead-end filtration is one of the simplest modes of operation for filters and hence requires minimum accessories such as tubing/piping, tanks, controls, and footprint.

Dead-end filters described using microporous membranes manufactured out of synthetic polymers such as polyethersulfonate, polyamide, cyanoacrylate, and polyvinylidene fluoride are used extensively for sterile processing. They are used for adding media to the bioreactor, bioburden reduction in cell harvest clarification, chromatography column protection, and final filtration of the purified bulk drug substance. These filters often come attached to disposable bags and are gamma sterilized.

The most common dead-end filtration devices are filter cartridges for reusable processes or capsules for fully disposable processes. They are used in wide-ranging applications as pre- and sterilizing-grade filters in upstream as well as downstream applications including media filtration, intermediate product pool filtrations, and in form, fill, and finish for the sterilization of drug substance. Dead-end filter devices are also used for sterilizing grade air and vent filtration for cell harvest and clarification, and, most recently, for viral clearance and membrane chromatography.

Cross-Flow Filtration

In chemical engineering, biochemical engineering, and protein purification, cross-flow filtration (also known as *tangential flow filtration*) is a type of filtration, a particular unit operation. Cross-flow filtration is different from dead-end filtration in which the feed is passed through a membrane or bed, the solids being trapped in the filter, and the filtrate being released at the other end. Cross-flow filtration gets its name because the majority of the feed flow travels tangentially across the surface of the filter, rather than into the filter. The principal advantage of this is that the filter cake (which can blind the filter) is substantially washed away during the filtration process, increasing the length of time that a filter unit can be operational. It can be a continuous process, unlike batch-wise dead-end filtration. This type of filtration is typically selected for feeds containing a high proportion of small-particle-size solids (where the permeate is of most value) because solid material can quickly block (blind) the filter surface with dead-end filtration. Industrial examples of this include the extraction of soluble antibiotics from fermentation liquors.

Since in cross-flow filtration the feed stream is led across or tangential to the filter material surface and is recycled continuously around the filter, this requires more complex equipment and controls, but the retentate is allowed to pass through the filter device multiple times by recirculation. Thus, it is possible to perform concentration or buffer-exchange processes. Additionally, for liquids with a heavy load of suspended particles, the filter is kept from clogging as the turbulent flow of the feed across the filter removes deposited materials, something that is not possible in dead-end filtration.

Filtration Media

Filter media generally comprises layers of solid materials in a network or mesh with voids, pores, and channels that allow the passage of liquid but retain larger particles, larger than the size of the openings, which may be in nanometers.

Depth filters use their entire depth to retain particulate on the basis of sieving compounded by adsorption effects unlike retentive filters where the filtered material is concentrated on the surface. The depth filter media dominate prefiltration and clarification applications because of the high solid mass that is generally required to be removed at this stage.

Sieving or size exclusion have more uniform pore sizes throughout the bed and are thus used to remove selective size of particles; these filters, mostly membrane types, are ideal as sterilizing filters, for example, the commonly use 0.22 μm filter to sterilize liquid. While the main mechanism of their operation is sieving, the chemical nature of these membranes makes them a good base to adsorb organic substances.

Depth filter are made of fibers that are spread out on a substrate to make a mesh just like making paper; special additives such as activated carbon, ceramic fibers, and other such specific components are embedded with the help of a binder to form the filter.

Sheet filters are also made like paper using milled cellulose fibers and may contain diatomaceous earth or perlite along with a binder to strengthen the filter.

One of the world's largest suppliers of these filters in biopharmaceutical manufacturing is Pall (Table 10.1).

Because of their thickness, the sheet filters provide a slow filtration option yet are extensively used for prefiltration

Microglass fibers are also used to filter media; these are nonwoven spun fibers of borosilicate glass whose web is strengthened by a binder allowing for a 3D structure of asymmetric voids as small as 0.2 μm to act as sterilizing filters.

Polypropylene and polyester fibers are also used by spinning from polymer melt and bonded by the polymer itself giving better chemical compatibility as no binder is added to them. These are always the preferred filters over polyamide and cellulose filters. The convention method of their manufacture leaves pore sizes 20–50 μm making them unsuitable for sterile filtration; a special blown process is used to reduce the pore size in the range of 5–50 μm; further spinning is needed to reduce the size further to 1–10 μm range.

TABLE 10.1

Pall Filter Offering (www.pall.com)

Filter	Use	Type
Seitz® K–Series Depth Filter Sheets	Active pharmaceutical ingredients, clarification and prefiltration, plasma fractionation	Sheet filters and sheet filter modules
Seitz® K–Series Depth Filter Sheets	Beer, bottled water, dairy, food, soft drinks, spirits, wine	Sheet filters and sheet filter modules
Seitz® P-Series Depth Filter Sheets	Biotechnology, clarification and prefiltration, plasma fractionation	Sheet filters and sheet filter modules
Seitz® T-Series Depth Filter Sheets	Prefiltration, production	Sheet filters and sheet filter modules
Seitz® Z-Series Depth Filter Sheets	Active pharmaceutical ingredients, clarification and prefiltration	Sheet filters and sheet filter modules
Supracap™ 100 Depth Filter Capsules	Active pharmaceutical ingredients, biotechnology, cell separation, clarification and prefiltration, plasma fractionation, scale-up/ process development, vaccines	Capsules, sheet filters, and sheet filter modules
SUPRAcap™ 200 Encapsulated Depth Filter Modules	Active pharmaceutical ingredients, biotechnology, cell separation, clarification and prefiltration, plasma fractionation, scale-up/ process development, vaccines	Capsules, sheet filters, and sheet filter modules
Supracap™ 60 Depth Filter Capsules	Active pharmaceutical ingredients, biotechnology, cell separation, clarification and prefiltration, plasma fractionation, scale-up/ process development, vaccines	Capsules, sheet filters, and sheet filter modules
SUPRAdisc™ Depth Filter Modules	Active pharmaceutical ingredients, biotechnology, cell separation, clarification and prefiltration, plasma fractionation, scale-up/ process development, vaccines	Sheet filters and sheet filter modules
SUPRAdisc™ Depth Filter Modules	Biofuels and biotechnology, chemicals	Sheet filters and sheet filter modules
SUPRAdisc™ HP Depth Filter Modules	Active pharmaceutical ingredients, biotechnology, cell separation, clarification and prefiltration, plasma fractionation, scale-up/ process development, vaccines	Sheet filters and sheet filter modules
SUPRAdisc™ II Depth Filter Modules	Beer, food, juice, spirits, wine	Sheet filters and sheet filter modules
SUPRAdisc™ II Modules	Active pharmaceutical ingredients, biotechnology, clarification and prefiltration, plasma fractionation, scale-up/process development, vaccines	Sheet filters and sheet filter modules

Continued

TABLE 10.1 (*Continued*)

Pall Filter Offering (www.pall.com)

Filter	Use	Type
SUPRApak™ Depth Filter Modules	Beer, spirits	Sheet filters and sheet filter modules
SUPRApak™ SW Series Modules	Beer, beer—corporate brewers, beer—microbreweries, food, soft drinks, spirits	Sheet filters and sheet filter modules
T-Series Depth Filter Sheets	Active pharmaceutical ingredients, biotechnology, cell separation, clarification and prefiltration, plasma fractionation, scale-up/ process development, vaccines	Sheet filters and sheet filter modules
T-Series Depth Filter Sheets	Biofuels and biotechnology, chemicals	Sheet filters and sheet filter modules
T-Series Membrane Cassettes	Active pharmaceutical ingredients, clarification and prefiltration	Sheet filters and sheet filter modules

Polymer Membranes

The history of membrane filters goes back to hundreds of years:

Year	Important Development
1748	Abbe Nollet—water diffuses from dilute to concentrated solution
1846	The first synthetic (or semisynthetic) polymer studied by Schoenbein and produced commercially in 1869
1855	Fick employed cellulose nitrate membrane in his classic study *Ueber Diffusion*
1866	Fick, Traube, artificial membranes (nitrocellulose)
1907	Bechhold, pore size control, "ultrafiltration"
1927	Sartorius company, membranes available commercially
1945	German scientists, methods for bacterial culturing
1957	USPH, officially accepts membrane procedure
1958	Sourirajan, first success in desalinating water

The main advances in membrane technology (1960–1980) began in 1960 with the invention of the first asymmetric integrally skinned cellulose acetate reverse osmosis (RO) membrane. This development simulated both commercial and academic interest, first in desalination by reverse osmosis, and then in other membrane applications and processes. During this period, significant progress was made in virtually every phase of membrane technology: applications, research tools, membrane formation processes, chemical and physical structures, configurations, and packaging.

Membrane Morphology	
Isotropic	Anisotropic
Dense	Dense/Selective Skin
Porous	Porous Skin

FIGURE 10.1
Basic membrane morphology.

Two basic morphologies of hollow fiber membrane are *isotropic* and *anisotropic* (Figure 10.1). Membrane separation is achieved by using these morphologies.

The anisotropic configuration is of special value. In the early 1960s, the development of anisotropic membranes exhibiting a dense, ultrathin skin on a porous structure provided a momentum to the progress of membrane separation technology. The semipermeability of the porous morphology is based essentially on the spatial cross section of the permeating species, that is, small molecules exhibit a higher permeability rate through the fiber wall. While the anisotropic morphology of the dense membrane, which exhibit the dense skin, is obtained through the solution-diffusion mechanism. The permeation species chemically interacts with the polymer matrix and selectively dissolves in it, resulting in diffusive mass transport along the chemical potential gradient, as was demonstrated in the pervaporation process.

Type of the membrane configuration is given in Figure 10.2.

Hollow fiber is one of the most popular membranes used in industries. It is because of its several beneficial features that make it attractive for those industries. Among them are

- *Modest energy requirement*: In the hollow fiber filtration process, no phase change is involved. Consequently, it needs no latent heat. This makes the hollow fiber membrane have the potential to replace some unit operations that consume heat, such as distillation or evaporation columns.

- *No waste products*: Since the basic principle of hollow fiber is filtration, it does not create any waste from its operation except for the unwanted component in the feed stream. This can help to decrease the cost of operation to handle the waste.

- *Large surface per unit volume*: Hollow fiber has a large membrane surface per module volume. Hence, the size of hollow fiber is smaller than other types of membrane but can give higher performance.

- *Flexible*: Hollow fiber is a flexible membrane; it can carry out the filtration by two ways; it is either is "inside-out" or "outside-in."

- *Low operation cost*: Hollow fiber needs low operational costs compared to other types of unit operation.

Modules		
Hollow Fiber-Capillary	**Plate and Frame**	**Spiral Wound**
• Very small diameter membranes (< 1mm) • Consist large number of membranes in a module and self supporting. • Density is about 600 to 1200 m^2/m^3. (for capillary membrane), up to 30000 m^2/m^3. (hollow fiber) • Size is smaller than other module for given performance capacity. • Process "inside-out", permeate is collected outside of membrane • Process "outside-in", permeate passes into membrane bore.	• Structure is simple and the membrane replacement easy. • Similar to filter press • Density is about 100 to 400 m^2/m^3. • Membranes is placed in a sandwich style with feed sides facing each other. • Feed flows from its sides and permeate comes out from the top and the bottom of the frame. • Membranes are held apart by a corrugated spacer.	• Is formed from a plate and frame sheet wrapped around a center collection pipe • Density is about 300 to 1000 m^2/m^3. • Its diameter can up to 40cm. • Feed flows axial on cylindrical module and permeate flow into the central pipe. • Features: 1. High pressure durability. 2. Compactness. 3. Low permeate pressure drop and membrane contamination. 4. Minimum concentration polarization.
• Ceramic membranes are usually monoliths of tubular capillaries. • Channel sizes are in millimeter range. • Becomes a module by attaching end fittings and a means of permeate collection. • Many monoliths are usually incorporated into one modular housing.	• Not self supporting and normally are inserted in other materials' tubes with diameter more than 10mm. • Density is not more than 300 m^2/m^3. • Features: 1. Can operate with simple pre-treatment of feed liquid. 2. Membranes replacement is easy. 3. Easy to be washed.	

FIGURE 10.2
Membrane configuration.

However, it also has some disadvantages that contribute to its application constraints. Among the disadvantages are

- *Membrane fouling*: Membrane fouling of hollow fiber is more frequent than other membranes due to its configuration. Contaminated feed will increase the rate of membrane fouling, especially for hollow fiber.
- *Expensive*: Hollow fiber is more expensive than other membranes that are available in the market. It is because its fabrication method and expense is higher than other membranes.
- *Lack of research*: Hollow fiber is a new technology and, so far, there has been less research done on it compared to other types of membranes. Hence, more research will be done on it in the future.

Hollow fiber made of polymer cannot be used on corrosive substances and in high-temperature conditions. Various types of membrane processes can be found in almost all of the literature references.

There is considerable confusion in the open literature as to the distinction between a few membrane separation processes, that is, the microfiltration (MF), ultrafiltration (UF), and reverse osmosis (RO). Occasionally, one will see it referred to by other names such as "hyperfiltration (HF)." In order to distinguish these separation processes clearly that RO has, the separation range of 0.0001 to 0.001 m (i.e., 1 to 10 Å) or < 300 mol wt. RO is a liquid-driven membrane process, with the RO membranes capable of passing water while rejecting microsolutes, such as salts or low-molecular-weight organics (<1000 Da). A pressure driving force (1 to 10 MPa) is needed to overcome the force of osmosis that cause the water to flow from dilute permeate to concentrated feed. The principle use of this membrane process is desalination, which shows its great advantage over the conventional technique of desalination, that is, ion exchange.

The biotechnology industry, which originated in the late 1970s, has become one of the emerging industries that draw the attention of the world, especially with the emergence of genetic engineering as a means of producing medically important proteins during the 1980s. Two of the major interest applications of membrane technology in the biotechnology industry is the separation and purification of the biochemical product, as is often known as downstream processing, and the membrane bioreactor, which was developed for the transformation of certain substrates by enzymes (i.e., biological catalysts).

Since its introduction in the 1970s, the membrane bioreactor has gained a lot of attention over the other conventional production processes concerning the possibility of high enzyme density and hence high space-time yields. Whereas downstream processing is usually based on discontinuously operated microfiltration, the membrane bioreactor is operated continuously and is equipped with UF membranes. Two types of bioreactor designs are possible: dissolved enzymes, (as in used with the production of l-alanine from pyrurate) or immobilized enzymes membrane.

Membrane science began emerging as an independent technology only in the mid-1970s, and its engineering concepts still are being defined. Many developments that initially evolved from government-sponsored fundamental studies are now successfully gaining the interest of the industries as membrane separation has emerged as a feasible technology.

Today, membrane polymers used in pharmaceutical processes include polyvinylidene fluoride (PVDF), expanded polytetrafluorethylene (ePTFE), polyethersulfone (PESu), polyamide (PA), cellulose acetate (CA), regenerated cellulose (RC), and mixed cellulose ester (MCE), a mixture of cellulose nitrate (CN) and CA. Membranes provide the highest retention efficiency or the smallest pore sizes; the microfiltration membrane pore sizes range from 10 to 0.1 µm; the ultrafiltration membranes have pore sizes from 0.1 µm to a few nanometers, making them even more suitable for virus filtration. The nanofilters are in the range of 50 nm and smaller and rated for the molecular weight they can retain in kilodaltons. Microfiltration membranes are suitable for prefiltration, others for sterilization.

Microfiltration Cross-Flow

The traditional process for adopting cross-flow filtration involves the use of dense ultrafiltration membranes for the purpose of performing a concentration and buffer exchange operation on the target molecule pre- and post-column chromatography. When the ultrafiltration media is replaced with a microporous microfiltration membrane, one has the option of performing a wide range of separations involving larger species, while usually operating at much lower pressures.

As in the case of the ultrafiltration membrane applications, a suitable cross-flow microfiltration process may involve the concentration and "washing" of a particulate material. For example, as an alternative to the use of centrifugation, one could aseptically recover cells from a cell culture process and proceed to concentrate and wash the cells to remove contaminating macrosolutes and replace the fluid with a solution suitable for freezing the cells. This same type of sequence could be used to aseptically process liposomes and other drug delivery emulsions.

A second widely practiced use of cross-flow microfiltration is also an alternative to centrifugation or normal flow filtration for the clarification of the target molecule or virus from cells and cell debris. Highly permeable microfiltration membranes allow a large target molecule to pass through the filter, while providing complete retention of the contaminating particulates in a single scaleable step.

The filter media needs a device to use; earlier devices were stainless steel holders for flat discs; pleated inserts to provide a larger surface area followed this, and these were enclosed in plastic containers. While the stainless steel devices could be autoclaved, the plastic components needed gamma radiation to sterilize.

Hollow fiber devices contain bundles of numerous hollow fiber membranes, and they are placed coaxially into a pipe-like perforated cage and sealed using a resin so that either one or both ends of the module are open to give access to the feed or allow exit to the filtrate or permeate. They can be operated from both sides, the tube or the shell side.

Hollow fiber devices are established, for example, in virus clearance by Asahi Kasei Medical Planova®:

> Planova filters, the world's first filters designed specifically for virus removal, significantly enhance virus safety in biotherapeutic drug products, such as biopharmaceuticals and plasma derivatives. They exhibit unparalleled performance in removing viruses, ranging from the human immunodeficiency virus (HIV) to parvovirus B19, while providing maximum product recovery.

BioOptimal MF-SL™

Designed specifically for use in cell culture clarification applications, BioOptimal MF-SL filters enable biopharmaceutical manufacturers to improve the efficiency and effectiveness of their protein harvest step.

TechniKrom™

Asahi Kasei Bioprocess, Inc., provides technologically advanced bioprocess equipment, products, and services to the biopharmaceutical, pharmaceutical, veterinary, and nutraceutical industries that permit the lowest cost of production and highest innate product quality, especially in cGMP-regulated environments. We help our clients implement true manufacturing science in their facilities to enable their achievement of these goals.

GE Healthcare

Process scale hollow fiber cartridges offered by GE Healthcare are provided in eight basic configurations covering a membrane area range of 0.92 to 28 m² (9.9 to 300 ft²) depending on the fiber internal diameter.

Spectrum

Spectrum disposable CellFlo hollow fiber membranes are specially designed for the gentle and efficient separation of whole cells in microfiltration, bioreactor perfusion, and culture harvest applications. Very similar to GE's circulatory system, CellFlo combines the advantages of tangential flow microfiltration (0.2 μm and 0.5 μm pores sizes) with larger hollow fiber flow channels (1 mm ID) to provide gentler efficacious microseparations without the risk of cell lysis. Other cell separation technologies have a higher risk of lysis resulting in ruptured cells and culture harvests contaminated with intracellular macromolecules. Perfectly suited for continuous cell perfusion and bacterial fermentation, CellFlo membranes isolate secreted proteins while eliminating spent media containing metabolic wastes and drawing in fresh nutrient-rich media. Consequently, CellFlo enables cultures to grow to a higher cell density with higher viability providing as much as a tenfold increase of daily production of secreted proteins.

Whether performing a cell perfusion or conducting a simple microfiltration, disposable CellFlo modules can either be autoclaved or purchased irradiated (irr) for quick sterile assembly. Spectrum offers disposable CellFlo membrane modules in the full range of MiniKros and KrosFlo sizes and surface areas for processing volumes ranging from 500 mL to 1,000 L. All CellFlo membranes have a 1 mm fiber inner diameter, available in 0.2 μm and 0.5 μm pore sizes. Also to be considered are the MaxCell, Spectrum Laboratories CellFlo®, and KrosFlo® module families.

Millipore has developed a derivative hollow fiber system, Pellicon, and here is how it compares with standard hollow fiber filters:

Hollow Fiber versus Pellicon 2—Summary

Feature	Hollow Fiber	Pellicon 2
Robustness and reliability	Fiber is prone to stress failure	Very robust
Pressure capability	Low	High
Membrane choices	1. Polyethersulfone	1. Polyethersulfone 2. Regenerated cellulose 3. PVDF (for MICRO filtration)
Flow rate required to operate TFF processes	High flow rate resulting in • High energy consumption • Large piping • Compromised concentration ratio • Increased demand for floor space	Low flow rate resulting in • Low energy consumption • Small piping • High concentration ratio • Compact system size
Linear scale-up	Compromised by differences in the length of the flow channel in laboratory versus process scale cartridges	Identical length of flow channel in all cartridge sizes to facilitate predictable scaling results
Consistency of retention relative to retentate channel design	Compromised consistency due to • Open flow channels, no internal mixing	High consistency due to • Built-in static mixer, efficient internal mixing
Consistency of retention relative to the retentate flow channel length	Compromised consistency due to • Long flow channels	High consistency due to • Very short flow channels

Flat sheets can be arranged in a stack in place of pleated or hollow fiber filter media. The sheets are sealed so that they leave channels open to allow feed to pass through.

Conclusion

From prefiltration to remove large sediments to harvesting bacterial cultures to removing viruses in the final stages of biological drug manufacturing, filters play one of the most important roles. There are dozens of companies specializing in specific filtration processes and giants dominate the market with their Millipore Pod® system, the Pall Stax® system, or Sartoclear® XL Drums from Sartorius-Stedim. They also keep bringing newer filters, housings, arrangements, and recently in an integrated setup in their disposable factories concept. The manufacturer is highly advised to consult the current literature on the suitability of the type of filters used. More often than not, the suppliers are more than willing to understand the process and make recommendations. Obviously, cost is a serious concern but, when it is realized that in a cGMP-manufacturing environment having qualified a particular filter or a housing for a unit process, it is not easy to switch over to another type of filter or another supplier, the selection of these filters becomes a serious concern.

11

Regulatory Compliance

The logic of validation allows us to move between the two limits of dogmatism and skepticism.

Paul Ricoeur

The medical device industry has matured into hundreds of regulatory approvals worldwide, and with that has come a keen understanding of biocompatibility, leachability, and the safety of the plastic materials used. The disposable systems gaining wide appeal in bioprocessing have to revert to regulatory opinions on medical devices since there are no current guidelines available from the FDA or EMEA on the evaluation of disposable manufacturing factories for biological drugs.

There is sufficient guidance available on the testing and validation of any measurement device used in cGMP conditions, ranging from calibration to IQ/OQ/PQ to CFR 21 Part 11 compliance when there is a central processing unit (CPU) involved.

To date, no major regulatory authority has approved any product manufactured using a disposable bioreactor. It is not because of any risk factors in the use of disposable systems but in the inability of the manufacturers to file these applications. Generally, a manufacturer will file a regulatory marketing authorization application using manufacturing systems that are capable of producing at least 10% of the final batch size that the manufacturer plans to make when the drug is approved. The manufacturer also wants to make sure that the larger batches would be easily scaled-up. When using hard-walled systems, this is relatively easy as the fermenters and bioreactors supplied by a large equipment manufacturer offer several sizes and most of these are readily scalable. When it comes to disposable bioreactor systems, the largest-size bioreactor that has become available is 2,000 L in size and even that is an expensive offering as most manufacturers have simply emulated the hard-walled systems by placing a liner within them. All of the other hardware and software remains the same, making it a much more expensive undertaking. Even if Big Pharma is willing to accept those costs, the maximum size of 2,000 L is too low to convince betting on this system. Recently, a Korean company installed a hard-walled animal cell bioreactor greater than 80,000 L; however, this type of sizing is most likely to become obsolete as cell lines become more productive. As an example, a company using a cell line qualified 20 years go would have to use at least 20 times a larger scale than if they were able to replace them with a newer line.

While for drugs such as insulin and other bacterial-expression drugs, the size of production will remain high, drugs made using cell cultures are most likely to be made in smaller-size bioreactors as the lines enter a productivity of 10 g/L.

Still, there remains a large market for smaller-volume drugs, clinical supplies, and more recently cell therapy that holds great promise for the disposable manufacturing industry to begin to show its promise in terms of approval marketing authorizations.

The requirements for regulatory submission using a disposable system are exactly the same as those for hard-walled manufacturing systems. The usual validation, calibration, and operational qualification of the equipment are required. However, this does create a challenge especially when it comes to managing characterization, robustness, and scalability. The fact is that the manufacturer must submit robust studies to prove that all leachables are identified and that they do not affect the quality of the product if present. Leachables apply to every component from the mixing bags to transfer tubes to filters, connectors, and storage devices. The manufacturer is also required to demonstrate the scalability of membrane chromatography, flow distribution, and device variability.

There is a belief that membranes are more variable than chromatography resins and that it is the manufacturer's responsibility to prove through extensive process analytic technology (PAT) practice that the variation does not affect the quality.

Finally, the manufacturer must demonstrate comparability of the product with the innovator product if it is a biogeneric product. This includes lot-to-lot variations as well, and this is where some problems can arise. There will never be two batches of a biological drug that are always the same; variability is inherent in biological systems and it is for this reason alone that regulatory agencies require strict environmental and compliance control.

To demonstrate how the FDA views the single-use or disposable technology, a review of the *Manual for Biologicals Compliance Program Guidance Manual*, Chapter 45—Biological Drug Products; Inspection of Biological Drug Products (CBER) 7345.848, implementation date of October 1, 2010, (www.fda.gov/cber/cpg/7345848.htm), has one entry for "single-use," and none for "disposable" or "plastic." The entry on "single-use" states as follows:

> **8c) Filtration** There are various types of filtration methods, such as diafiltration, ultrafiltration, and microfiltration that may be used in the purification of vaccine products. Some of the filters used may be single-use and some may be multiuse. The filters are usually placed within a filter housing apparatus. The criteria used for the evaluation of the column purification should also be applied to the filter housings and the multi-use filters.

The concerns about leachability or extractability remain to be challenged in filed applications and, as with any new event, the industry is in a waiting mode to see how the FDA or EMEA react to these submissions.

Regulatory Barriers

There are no extraordinary regulatory barriers in the deployment of disposable systems except the need to validate leachables. Not surprisingly, even the hard-walled manufacturing systems inevitably use plastic for such uses as filtration, packaging, and storage. Whether these devices or components are steam cleaned or not is irrelevant. So, there is already a precedent to show the safety of plastics in manufacturing.

The use of disposables is supposed to reduce regulatory barriers substantially. It has been clearly established that if the manufacturer does not have to do any cleaning validation between batches, the cost will be reduced. There are other features of a facility using disposables, including how this can reduce regulatory barriers that should be considered.

One of the easiest and perhaps most robust means of assuring lack of cross-contamination is to dedicate a facility for a single molecule; this way only one cell line enters the facility and only one molecule goes out; all equipment is dedicated, and nothing in the facility is shared with another operation by the manufacturer. The cleaning between batches remains a task, but its importance is significantly reduced as no new contaminants are expected. Dedicating a facility to one molecule is not possible today for most companies unless they happen to be the innovator producing very large quantities of the product. However, the smaller footprint required in a disposable facility can make it possible to dedicate facilities to individual molecules. First, the elimination of sterilization-in-place (SIP)/cleaning-in-place (CIP) reduces capital cost substantially, and the requirement of the total space needed is also reduced substantially. Combining this with adopting newer techniques in producing purified water (e.g., double reverse osmosis (RO)/ electrodeionization (EDI), the company can completely eliminate the large engineering infrastructure needed to provide water in the manufacturing area. It is projected that almost 90% of the water used in traditional facilities is for CIP/SIP and autoclave use. Amgen uses 80,000 gallon of water per day in its Rhode Island facility, most of it for CIP/SIP. The cost of water is very high; use of disposables brings the first savings in terms of energy cost reduction (required in water distillation) and the capital costs of distillation and storage. The new RO systems do not have storage vessels and work on an on-demand basis.

Another design element that reduces the footprint of a manufacturing facility using disposables is the size of equipment installed; while many suppliers are still following the traditional design of a 3D stainless steel bioreactor with a liner, the real breakthrough in the field comes from 2D bioreactors that do not need an outer container and can be used horizontally, not vertically like most 3D bioreactors. Laying down the bioreactors horizontally makes the facility design compact and allows the use of smaller ceiling heights, which adds substantially to cost savings in the facility design.

A model facility that makes the maximum utilization of disposable systems was recently constructed in Chicago. Figure 11.1 shows the layout of the facility to manufacture erythropoietin and granulocylte-colony stimulating factor (G-CSF), about 1 kg per year. The total square footage including the laboratory is about 12,000 square feet; another facility that produced two monoclonal antibodies, about 100 kg per year, required about 20,000 square feet total for manufacturing operations (see Figure 11.2).

Such compact facility designs allow manufacturers to dedicate facilities or independent suites to each molecule as done by the Chicago biotechnology company, Therapeutic Proteins, Inc. This reduced their regulatory barriers significantly. Similar facilities using a traditional hard-walled design would take up at least three times more space and would cost at least five times more to construct and furnish.

A comparison of the cost of manufacturing of biological APIs using disposable systems is given in Table 11.1.

The cost reduction shown earlier represents the most conservative calculation; with appropriate adjustments of product flow, validation cycles, and batch sizes, it is anticipated that the cost of production in well-coordinated disposable systems should not be more than 50% of the traditional cost.

It is now possible for contract manufacturing companies to establish isolated production areas for at least one type of cell line; keeping animal cell lines in one suite would significantly reduce their regulatory costs of compliance.

A summary of pros and cons of the two systems is given in Table 11.2.

Irradiation and Sterilization Validation

In many cases, microbial control or sterility is required to ensure product purity and safety. Radiation sterilization is a common means of microbial control and sterilization applied to single-use systems. The standard methods for validating radiation sterilization are often not clearly understood since the industry has mostly operated on using steam sterilization; however, a keen understanding of how irradiated components are validated is important for regulatory filing. The standards are established by the American National Standards Institute (ANSI), the Association for the Advancement

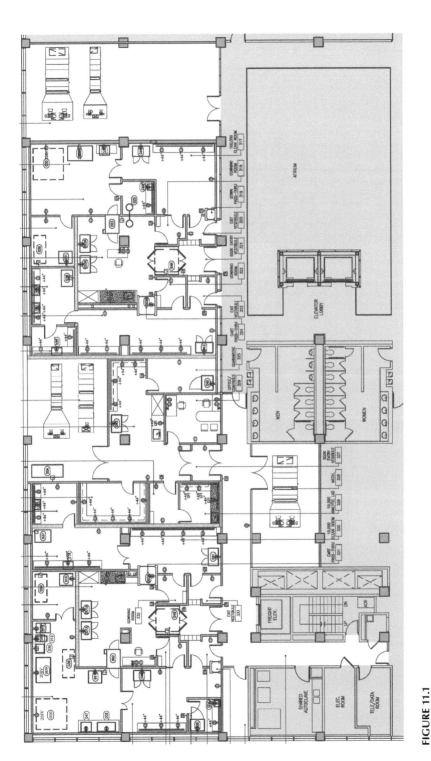

FIGURE 11.1
Layout of a manufacturing facility producing 1 kg each of erythropoietin and G-CSF in Chicago; total square footage: 12,000 ft².

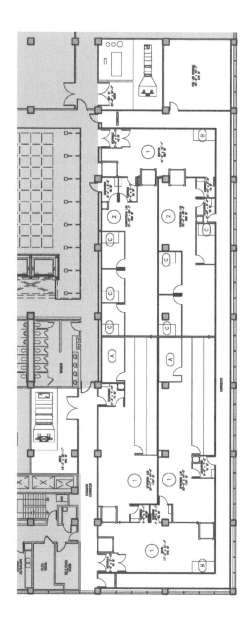

FIGURE 11.2
Layout of a monoclonal antibody manufacturing facility producing two molecules, 100 kg each per year, in Chicago; total square footage: 20,000 ft².

TABLE 11.1

Unitary Comparative Cost of Manufacturing of a
Biological API Using Stainless Steel and
Disposable Systems

Category	Stainless Steel	Disposable
Capital charge	37	10
Materials	14	11
Media	3	3
Buffer	1	1
CIP	1	1
QC tests	9	6
Consumables	11	14
Resins/MA	4	4
Bags, disposables	0	3
Filters	7	7
Labor	29	17
Process	11	10
Quality	14	5
Indirect	4	2
Other	10	5
Insurance and other	2	2
Maintenance	2	1
Utilities	6	2
TOTAL	100	57

Note: The capital cost is the amortization assuming the
facility is built in Europe or the United States.

Source: Data for stainless steel from Sinclair, A., How
geography affects the cost of biomanufacturing,
BioProcess International, June 2010, 516–519.

of Medical Instrumentation (AAMI), the International Organization for
Standardization (ISO), and ASTM International (formerly the American
Society for Testing and Materials). Validating the sterility of an irradiated
bag can be a difficult and expensive process.

Gamma irradiation is the application of electromagnetic radiation (gamma
rays) emitted from radionuclides such as Cobalt 60 (60 Co) and Cesium 137
(137 Cs) isotopes. Gamma rays are not retarded by most materials and can
penetrate through most disposable bioprocess system components. Living
organisms are inactivated by damage to their nucleic acids resulting from
this ionizing irradiation. Gamma rays are also not retained by material and
leave no residual radioactivity.

Gamma irradiation dosage is measured in kilogray (kGy) units, which
quantify the absorbed energy of radiation. One gray is the absorption of 1 J of
radiation energy by 1 kg of matter (1 kGy = 1 J/gm). Dosages ≥8 kGy are generally adequate to eliminate low bioburden level. In cases where bioburden

TABLE 11.2

Comparison of Pros and Cons of Disposable Versus Stainless Steel Systems

Disposable	Stainless Steel
Pros	
Presterile, ready to use	Proven technology
Easy setup	Scalability virtually unlimited
Eliminates CIP/SIP	
Low capital outlay	
Low validation requirements	
Increased flexibility	
Cons	
New technology	Hard to clean and maintain
Volume limitations	Large capital investment
	Extensive CIP/SIP cycle
	Extensive validation
	Expensive installation
	Accessories also of stainless steel

level is elevated (>1,000 colony-forming units, or cfu, per unit), as may occur with very large single-use systems, higher doses may be required to achieve sterility.

Sterility assurance level (SAL) is a term used in microbiology to describe the probability of a single unit being nonsterile after it has been subjected to the sterilization process. For example, medical device manufacturers design their sterilization processes for an extremely low SAL— "one in a million" devices should be nonsterile. SAL is also used to describe the killing efficacy of a sterilization process, where a very effective sterilization process has a very low SAL.

In microbiology, it is impossible to prove that all organisms have been destroyed because (1) they could be present but undetectable simply because they are not being incubated in their preferred environment, and (2) they could be present but undetectable because their existence has never been discovered. Therefore, SALs are used to describe the probability that a given sterilization process has not destroyed all of the microorganisms.

Mathematically, SALs referring to probability are usually very small numbers and so are properly expressed as negative exponents ("the SAL of this process is 10 to −6"). SALs referring to the sterilization efficacy are usually much larger numbers and so are properly expressed as positive exponents ("the SAL of this process is 10 to 6"). In this usage, the negative effect of the process is sometimes inferred by using the word *reduction* ("this process gives a six-log reduction"). Because of this ambiguity, group discussions of SAL must define the terminology before setting standards.

SALs can be used to describe the microbial population that was destroyed by the sterilization process. Each log reduction (10^{-1}) represents a 90% reduction in microbial population. So, a process shown to achieve a "6-log reduction" (10^{-6}) will reduce a population from a million organisms (10^6) to very close to zero, theoretically. It is common to employ overkill cycles to provide the greatest assurance of sterility for critical products such as implantable devices.

SALs describing the *"Probability of a Nonsterile Unit"* (PNSU) are expressed more specifically as PNSU in some literature.

Generally, 25 kGy can achieve sterility with a SAL of 10^{-6}. Even with elevated bioburden levels, bioburden reduction can be achieved with lower probabilities of sterility (e.g., SAL of 10^{-5} or 10^{-4}). Products irradiated to such SALs are still sterile but have higher probabilities of nonsterility and may not meet standards for validated sterile claims as specified in industry standards for sterilization of health care products.

Gamma irradiation also causes ionization and excitation of polymer molecules. Some polymers show higher resistance to irradiation-induced changes than others; all polymers are affected to some degree. Repeated irradiation of single-use systems or components should be avoided, as the effect is cumulative.

The concepts and protocols described in the current industry standards for sterilization of health care products by gamma irradiation are applicable to disposable bioprocess systems as well. There is the requirement of the validation of the efficacy and reproducibility of the sterilization process, based on determination of average bioburden and subsequent sterility testing of systems after minimal radiation dose exposures. Systems validated as sterile are also subject to routine audits involving bioburden and sterility testing. Components or systems requiring zero or low bioburden when applied in nonsterile processes do not require a validated sterile claim and may be qualified as microbially controlled.

It is important to establish where a full validation of the container is required. Cell culture bioprocess is divided into several processing stages: upstream, harvesting, cell separation, depth filtration, and membrane filtration. In the downstream stage, the target molecule is subjected to a series of separation, purification, and concentration stages applying chromatography and membrane filtration to ultimately produce a purified bulk drug product. This is followed by a final formulation stage where the purified bulk (API or biological) is transformed to a stable formulation and sterilized by filtration (or filled aseptically in sterile containers). All stages where the process is claimed to be sterile would require a validation but not to processes listed as "microbially controlled," even though they may have low or no burden. It is obvious that these stages of preparation of biological drugs cannot be done under sterile conditions and thus generally do not require sterility validation, and the manufacturer should not claim the process as sterile and instead they should be listed as operated under a high degree of microbial control, as can be provided by gamma irradiation without sterilization validation. This is important to reduce the time burden of validation.

In upstream processing, bacterial (e.g., *Escherichia coli*) cell cultures tend to be operated in short time frames and are fairly resistant to overgrowth, because of genetic coding of antibiotic resistance and the media contains these antibiotics, by low levels of contaminant bacteria. It may be prudent to classify the bioreactor as microbially controlled and not sterile, allowing the manufacturer to not perform any validation of gamma-sterilized systems. This may not, however, be the case when operating mammalian cell cultures that run for a longer time, weeks at a time. Here, the bacterial contamination can be serious, and sterilization validation may be required and validation necessitates complying with regulatory compliance.

Cell harvesting and downstream processing steps are rarely validated for sterility due to complexities and/or limitations of their equipment, and especially during process development. The process equipment is generally chemically disinfected or sanitized and maintained as microbially controlled for zero or low bioburden. Intermediate holds are kept at low temperature to prevent microbial contamination. In cases where these process steps are not claimed to be sterile, it is unnecessary to validate the sterility of single-use systems for preparation of process buffer feed solutions or intermediate holds. Buffer solutions or intermediates are filtered through irradiated bioburden reduction filter systems, as microbially controlled feeds are generally suitable for process steps not validated as sterile.

Even at the stage of final formulation and filling, it is not necessary to claim sterile processing as a nonsterile finished bulk API is subsequently processed with sterilization. The only steps requiring disposable systems to be validated as sterile are in the preparation of sterile API and the aseptic filling of sterile containers; that is, if the fluid is to be claimed as sterile, then the single-use system it is filled into must have that validated claim.

Several industry standards are used for sterilization validation of gamma-irradiated health-care products. The three parts of ANSI/AAMI/ISO 11137:2006 are

- Part 1: Requirements for Development, Validation, and Routine Control of a Sterilization Process for Medical Devices specifies requirements for development, validation, process control, and routine monitoring in the radiation sterilization for health-care products. Part 1 applies to continuous and batch-type gamma irradiators using the radionucleotides 60 Co or 137 Cs, and to irradiators using a beam from an electron or x-ray generator.

- Part 2: Establishing the Sterilization Dose describes methods that can be used to determine the minimum dose necessary to achieve the specified requirement for sterility, including methods to substantiate 15 or 25 kGy as the sterilization dose.

- Part 3: Guidance on Dosimetric Aspects provides guidance on dosimetry for radiation sterilization of health-care products and dosimetric aspects of establishing the maximum dose (product qualification); establishing the sterilization dose; installation qualification; operational qualification; and performance qualification.

ANSI/AAMI/ISO standards describe two methodologies for sterilization validation:

- Dose setting methods (Method 1 or Method 2, ANSI/AAMI/ISO 11137), which take into account distribution and radiation resistance of product bioburden, and to a limited extent, the end use of the product
- Dose substantiation methods (VDmax methods), which entail experimentation designed to qualify predetermined gamma dosages as a sterilization dose (1–3)

Table 11.3 summarizes the similarities and differences of the various methods.

Sterility testing of bioprocess systems or components presents significant challenges because of their large size that may not fit ordinary biosafety cabinets for incubation testing. There is also a problem in assuring that the outer surface does not contaminate the inner surface. Several common approaches can lessen the difficulties posed by these situations. The Sample Item Portion (SIP) approach allows for manufacturing and testing a reduced-scale product. SIP multipliers can be determined through experimentation or simply by taking a ratio of the model to the largest system manufactured. The Fluid Path (FP) testing may be simpler than testing an entire product because a product acts as its own barrier to contamination. This entails partially filling the product with a sterile buffer to ensure that all surfaces are wetted, then agitating the article by hand to promote suspension of organisms into the buffer. The buffer then is removed and tested for bioburden load through standard microbiological methods. To aid in handling, a large article can be split into parts that are more easily manipulated in the laboratory before testing or irradiation.

TABLE 11.3

Comparing Standard Irradiation Sterilization Methods for Healthcare Products

	ISO 11137 Method 1	ISO 11137 Method 2A	ISO 11137 Method 2B	ISO 11137 Method VDmax (25 kGy)	ISO 11137 Method VDmax (15 kGy)	AAMI TIR33 Method VDmax(15–35 kGy)
Limit CFU/unit (average)	0.1 to 0.9 or					
≥1 up to 1,000,000	None	Low	≤1,000	≤1.5	≤0.1 up to 440,000	Ranging from 15 to 35 kGy
Qualified sterilization manufacturing dose	ETE	ETE	ETE	25 kGy	15 kGy	
Units for initial bioburden determination	≥30 (≥10 units from three batches)	NA	NA	≥30 (≥10 units from three batches)	≥30 (≥10 units from three batches)	≥30 (≥10 units from three batches)
Units for initial verification dose	100					
180 units = 20 units at nine doses		≥280				
+100 units per verification run	≥260					
160 units = 20 units at eight doses						
+100 units per verification run	≥10	≥10	≥10			
Units for dose audit bioburden	≥10	≥10 units for routine monitoring	≥10 units for routine monitoring	≥10	≥10	≥10
Units for dose audit sterilization	100	100	100	≥10	≥10	≥10
SAL achieved	UD(10^{-2} to 10^{-6} range supported)	UD	UD	10^{-6}	10^{-6}	10^{-6}

Note: ETE = Established through experimentation; UD = user defined; NA = not applicable.

12

Environmental Concerns

The environment is everything that isn't me.

Albert Einstein

The carbon footprint of disposable technologies is larger than that of the reusable systems producing more solid waste; however, this must be studied in the light of the overall impact and not just in isolation.

The solid waste in disposable systems is mainly used plastic components, from bioreactor bags to connectors to filters. Note that even the stainless steel industry produces substantial solid waste, which includes plastic waste. The waste is of two types: one that can be folded and readily compressed and the other that is hard to compress, such as filter capsules and cartridges. It is important to understand that the size of solid waste produced in a disposable system reduces as a percentage of total waste as the batch sizes become larger. This is because some of the basic components are not related to the size of the batch. A rule of thumb to observe is that the size of solid wastes is about 10%–12% of the total batch size, so a batch of 5,000 L would produce waste of about 500 kg.

Disposal of solid waste from manufacturing can be a cumbersome task if the types of waste are different in terms of biosafety decontamination requirements. It is important that the components to be discarded be identified with the hazard before they are put to use.

Biosafety

How waste is handled depends to a large degree on its biosafety status. Since all disposable components that come in contact with a GMO may have to be equally treated depending on their biosafety status, a good understanding of the NIH Guidelines for Research Involving Recombinant DNA Molecules should be reviewed; the most recent version was issued in January 2011.

Since the bioprocessing industry is concerned mainly about two types of host cells, animal tissue such as CHO cells and bacterial organisms such as *E. Coli*, there the discussion pertains to both of these selections. An appendix to this chapter provides Appendix K of the NIH Guidelines for future reference. In the first instance, NIH classifies the hazard into the following four categories.

Risk Group 1 (RG1)	Agents that are not associated with disease in healthy adult humans
Risk Group 2 (RG2)	Agents that are associated with human disease that is rarely serious and for which preventive or therapeutic interventions are *often* available
Risk Group 3 (RG3)	Agents that are associated with serious or lethal human disease for which preventive or therapeutic interventions *may be* available (high individual risk but low community risk)
Risk Group 4 (RG4)	Agents that are likely to cause serious or lethal human disease for which preventive or therapeutic interventions are *not usually* available (high individual risk and high community risk)

Most of the GMOs that are used would fall in Risk Group 1 or 2. There is also an exempt group that is provided in Appendix C of the NIH Guidelines.

Appendix C-VIII-E i.e., the total of all genomes within a Family shall not exceed one-half of the genome.

Appendix C-I (*Recombinant DNA in Tissue Culture*) states: Recombinant DNA molecules containing less than one-half of any eukaryotic viral genome (all viruses from a single family being considered identical— see Appendix C-VIII-E, *Footnotes and References of Appendix C*), that are propagated and maintained in cells in tissue culture are exempt from these *NIH Guidelines* with the exceptions listed in Appendix C-I-A.

Appendix C-I-A. Exceptions

The following categories are not exempt from the NIH Guidelines: (i) experiments described in Section III-A which require Institutional Biosafety Committee approval, RAC review, and NIH Director approval before initiation, (ii) experiments described in Section III-B which require NIH/OBA and Institutional Biosafety Committee approval before initiation, (iii) experiments involving DNA from Risk Groups 3, 4, or restricted organisms (see Appendix B, Classification of Human Etiologic Agents on the Basis of Hazard, and Sections V-G and V-L, *Footnotes and References of Sections I through IV*) or cells known to be infected with these agents, (iv) experiments involving the deliberate introduction of genes coding for the biosynthesis of molecules that are toxic for vertebrates (see Appendix F, *Containment Conditions for Cloning of Genes Coding for the Biosynthesis of Molecules Toxic for Vertebrates*), and (v) whole plants regenerated from plant cells and tissue cultures are covered by the exemption provided they remain axenic cultures even though they differentiate into embryonic tissue and regenerate into plantlets.

Other exemptions may also apply to bacteria and yeast but with the following limitations:

Appendix C-II. *Escherichia coli* K-12 Host–Vector Systems

Experiments which use *Escherichia coli* K-12 host–vector systems, with the exception of those experiments listed in Appendix C-II-A, are exempt from the NIH Guidelines provided that: (i) the *Escherichia coli* host does not contain conjugation proficient plasmids or generalized transducing phages; or (ii) lambda or lambdoid or Ff bacteriophages or nonconjugative plasmids (see Appendix C-VIII. *Footnotes and References of Appendix C*) shall be used as vectors. However, experiments involving the insertion into *Escherichia coli* K-12 of DNA from prokaryotes that exchange genetic information (see Appendix C-VIII. *Footnotes and References of Appendix C*) with *Escherichia coli* may be performed with any *Escherichia coli* K-12 vector (e.g., conjugative plasmid). When a nonconjugative vector is used, the *Escherichia coli* K-12 host may contain conjugation-proficient plasmids either autonomous or integrated, or generalized transducing phages. For these exempt laboratory experiments, Biosafety Level (BL) 1 physical containment conditions are recommended. For large-scale fermentation experiments, the appropriate physical containment conditions need be no greater than those for the host organism unmodified by recombinant DNA techniques; the Institutional Biosafety Committee can specify higher containment if deemed necessary.

Appendix C-II-A. Exceptions

The following categories are not exempt from the NIH Guidelines: (i) experiments described in Section III-A which require Institutional Biosafety Committee approval, RAC review, and NIH Director approval before initiation, (ii) experiments described in Section III-B which require NIH/OBA and Institutional Biosafety Committee approval before initiation, (iii) experiments involving DNA from Risk Groups 3, 4, or restricted organisms (see Appendix B, Classification of Human Etiologic Agents on the Basis of Hazard, and Sections V-G and V-L, Footnotes and References of Sections I through IV) or cells known to be infected with these agents may be conducted under containment conditions specified in Section III-D-2 with prior Institutional Biosafety Committee review and approval, (iv) large-scale experiments (e.g., more than 10 liters of culture), and (v) experiments involving the cloning of toxin molecule genes coding for the biosynthesis of molecules toxic for vertebrates (see Appendix F, *Containment Conditions for Cloning of Genes Coding for the Biosynthesis of Molecules Toxic for Vertebrates*).

Appendix C-III. Saccharomyces Host–Vector Systems

Experiments involving *Saccharomyces cerevisiae* and *Saccharomyces uvarum* host–vector systems, with the exception of experiments listed in Appendix C-III-A, are exempt from the NIH Guidelines. For these exempt experiments, BL1 physical containment is recommended. For large-scale fermentation experiments, the appropriate physical containment conditions need be no greater than those for the host organism unmodified by recombinant DNA techniques; the Institutional Biosafety Committee can specify higher containment if deemed necessary.

Appendix C-III-A. Exceptions

The following categories are not exempt from the NIH Guidelines: (i) experiments described in Section III-A which require Institutional Biosafety Committee approval, RAC review, and NIH Director approval before initiation, (ii) experiments described in Section III-B which require NIH/OBA and Institutional Biosafety Committee approval before initiation, (iii) experiments involving DNA from Risk Groups 3, 4, or restricted organisms (see Appendix B, *Classification of Human Etiologic Agents on the Basis of Hazard*, and Sections V-G and V-L, *Footnotes and References of Sections I through IV*) or cells known to be infected with these agents may be conducted under containment conditions specified in Section III-D-2 with prior Institutional Biosafety Committee review and approval, (iv) large-scale experiments (e.g., more than 10 liters of culture), and (v) experiments involving the deliberate cloning of genes coding for the biosynthesis of molecules toxic for vertebrates (see Appendix F, *Containment Conditions for Cloning of Genes Coding for the Biosynthesis of Molecules Toxic for Vertebrates*).

Appendix C-IV. *Bacillus subtilis* or *Bacillus licheniformis* Host–Vector Systems

Any asporogenic *Bacillus subtilis* or asporogenic *Bacillus licheniformis* strain which does not revert to a spore-former with a frequency greater than 10-7 may be used for cloning DNA with the exception of those experiments listed in Appendix C-IV-A, *Exceptions*. For these exempt laboratory experiments, BL1 physical containment conditions are recommended. For large-scale fermentation experiments, the appropriate physical containment conditions need be no greater than those for the host organism unmodified by recombinant DNA techniques; the Institutional Biosafety Committee can specify higher containment if deemed necessary.

Appendix C-IV-A. Exceptions

The following categories are not exempt from the NIH Guidelines: (i) experiments described in Section III-A which require Institutional Biosafety Committee approval, RAC review, and NIH Director approval before initiation, (ii) experiments described in Section III-B which require NIH/OBA and Institutional Biosafety Committee approval before initiation, (iii) experiments involving DNA from Risk Groups 3, 4, or restricted organisms (see Appendix B, *Classification of Human Etiologic Agents on the Basis of Hazard*, and Sections V-G and V-L, *Footnotes and References of Sections I through IV*) or cells known to be infected with these agents may be conducted under containment conditions specified in Section III-D-2 with prior Institutional Biosafety Committee review and approval, (iv) large-scale experiments (e.g., more than 10 liters of culture), and (v) experiments involving the deliberate cloning of genes coding for the biosynthesis of molecules toxic for vertebrates (see Appendix F, *Containment Conditions for Cloning of Genes Coding for the Biosynthesis of Molecules Toxic for Vertebrates*).

In addition to those exemptions, the NIH distinguishes laboratory use from Good Large-Scale Practice (GLSP); the latter applies to manufacturers and makes exceptions from various compliances required in a laboratory.

In summary, in the United States, mammalian cells used for the production of recombinant proteins normally require BL1 level in laboratory and a GLSP level at large-scale commercial production, which means there are special containment or decontamination requirements and disposable bioreactors can be discarded along with other solid waste. When using bacteria, this may not be the case and usual procedures for biodecontamination would apply unless they fall under IIIC exemption category. Since most of the future products are likely to be produced in CHO cells such as the mAbs, it makes the cost of disposing of disposal components easily affordable.

However, discarding in this case would involve either incinerating them or taking them to a landfill as they are less likely to be recycled, according to the general consensus of the industry, though there is no clear reason why this is so. The recycling processes are extremely invasive and should remove any contaminants.

Those components that are readily recyclable are classified into the following categories:

Grade A = Recyclable plastic. Pure fractions of identified plastics. Examples: Preparation bags, cell culture bags, hold bags, cartridge bodies, single tubing, tank liners, and packaging material. Currently, grade A wastes constitute a small percentage, less than 10% of a complete disposable production chain. If designed appropriately, single components facilitate source separation of different grade A material batches (polyethylene [PE], polypropylene [Pp], polysulfone [PS]). Only grade A wastes are suitable for material recycling.

Grade B = mixed fractions of different plastics or multilayer films comprising polymers or thermosets. Examples: Most connectors, transfer systems, manifolds, tripolymer film bags. complete filtration cartridges, and most bag bioreactors. Up to 95% of the disposable components of a single-use biomanufacturing chain is grade B material. This fraction is suitable for energy recovery but not for material recycling.

Grade C = fractions with a significant amount (>5%) of nonplastic material such as glass, ceramics, metals, and electronic components. Examples: cell culture bags with sealed-in sensors, pumps and pump heads, centrifuge cartridges, mixing systems, filtration cells, downstream processing (DSP) units, and disposable sensors. Grade C waste usually constitutes only a minor but increasing fraction generally less than 20% of total single use system (SUS) waste. If recycled or used for fuel production, grade C waste requires pretreatment (fractionating of plastics, metal removal). It is mandatory to separate electronics and sensors from bulk wastes in the European Union (EU) and in Switzerland to facilitate collection and recycling of this material.

Liquid Waste

Disposable factories produce very small quantity of liquid waste, except for used media, eluant from TFF, and downstream columns. The combined liquid wastes for complete manufacturing chains typically contain only 2%–5% loading of CIP agents (caustic, acids) compared with SS systems.

Incineration

According to the U.S. Environmental Protection Agency, "incineration is a widely-accepted waste treatment option with many benefits. Combustion reduces the volume of waste that must be disposed in landfills, and can reduce the toxicity of waste." Incineration is a method of disposal that is used in many countries, and some companies incinerate as part of their standard disposal policy. In the European Union, a number of directives specifically address the issue of waste incineration and disposal:

- Directive 2000/76/EC of the European Parliament and the Council on the incineration of waste.
- Directive 94/67/EC of the Council of the European Communities excludes incinerators for infectious clinical waste unless rendered hazardous according to Directive 91/689/EEC of the Council of the European Communities on hazardous waste.

In some cases, incineration also can result in significant energy recovery as discussed below.

Cogeneration is a process in which a facility uses its waste energy to produce heat or electricity. Cogeneration is considered more environmentally friendly than exhausting incinerator heat and emissions directly up a smokestack.

One bioprocess company sends all its waste to a facility that incinerates and uses it to generate electricity for a major U.S. city. Another company uses a waste heat boiler to make low-pressure steam. Although there are wide variations, the heat value of mixed plastics waste is estimated to be about 15,000 to 20,000 BTUs/lb (34,890 to 46,520 kJ/kg), which compares favorably to coal at 9,000 to 12,000 BTUs/lb (20,934 to 27,912 kJ/kg).

Cogeneration is more widely applied in Europe and Asia than in the United States. In the United States, this process is being installed increasingly at universities, hospitals, and housing complexes for which boilers and chillers can serve multiple large buildings. In the European Union, the European Parliament and Council Directive 94/62/EC on packaging and packaging waste, Article 6 addresses energy recovery by incineration, and Article 10 addresses standardization.

In addition, the standard EN 13431 Packaging—Requirements for Packaging Recoverable in the Form of Energy Recovery, Including Specification of Minimum Inferior Calorific Value specifies requirements for packaging to be classified as recoverable in the form of energy and sets out procedures for assessing conformity with those requirements. The scope is limited to factors under a supplier's control.

Pyrolysis

Pyrolysis is a method for converting oil from plastics such as PE, PP, and polystyrene (PS) that can be used as fuel for internal combustion engines, generators, boilers, and industrial burners. Plastics are separated into oil, gas, and char residue by being heated in a pyrolysis chamber. Gas flowing through a catalytic converter is converted into a distillate fraction by the catalytic cracking process (enzymatically breaking the complex molecules down into simpler ones). The distillate is cooled as it passes through a condenser and then is collected in a recovery tank. From the recovery tank, the product is run through a centrifuge to remove contaminates. The clean distillate then is pumped to a storage tank.

About 950 mL of oil can be recovered from 1 kg of certain types of plastics. A comparison of the distillate produced by pyrolysis and regular diesel shows good similarity between the fuels, with the advantage that distillate from pyrolysis burns cleaner.

In-house incineration has, however, not been widely adopted as a means of minimizing solid waste in biopharmaceutical manufacture. With rising disposal costs, this may change rapidly; in most instances, this would be sourced out.

Landfilling of plastic wastes is expected to decrease in importance as a disposal option for solid wastes. The supposed advantages of landfills such as low operating and capital costs, high local availability, and energy production may no longer realistically apply, if indeed they ever did. Current landfilling practices include both direct disposal of nonhazardous waste and disposal of hazardous wastes after pretreatment.

Grind and Autoclave

All materials that have been in contact with biopharmaceutical components or with bioagents must be regarded as hazardous. Typical pretreatment for hazardous wastes includes grinding and autoclaving, as is a common practice with hospital waste. Some items are pretreated and shredded before landfilling. This option is appealing because it may be accepted as safe in some cases and reduces landfill volume compared with unshredded product. Additional discussions are ongoing regarding use of other hospital waste treatments such as autoclaving, thereby making a single-use system or component suitable for disposal in a standard municipal waste incinerator or landfill (if allowed). Some companies dispose of their used components or systems into a grinder–autoclave currently used at many hospitals, the National Cancer Institute, and the National Institutes of Health, among others.

This combination of mechanical and physical pretreatment significantly reduces the amount of waste for disposal, which can be reduced still further if it is then mechanically compacted. Furthermore, high-temperature pretreatment at, for example, 75°C, will almost completely inactivate biological contamination and destroy most (but not all) pharmaceutical contaminants as well. Higher temperature treatment up to 130°C, or gamma-ray irradiation, can ensure complete destruction of all temperature-resistant contaminants. One potential drawback of autoclaving is the possible production of leachable by-products from the plastics or biopharmaceutical components, which can also result from predecontamination with chlorine dioxide, leading to a higher risk of soil and water contamination through landfill leakage. The use of chlorine as a disinfectant is not regarded as environmentally sound because it can lead to undesirable atmospheric emissions. Wastes that are strongly acidic or caustic through contamination with pH control or CIP agents must be neutralized before landfilling.

Untreated, and ground and autoclaved (or otherwise decontaminated) waste exhibits high long-term stability in landfills and is not susceptible to biodegradation. As a consequence, no methane is produced in this fraction of landfilled waste. As energy is required for waste pretreatment and transport, the overall energy balance for waste disposed in a landfill is negative.

Landfill

Some companies choose landfilling. Its potential as an option varies based on municipal and regional regulations as well as on a product's use before disposal.

> Untreated: One industry player landfills an untreated component because its system does not require prior treatment. Another company whose products do not require landfill pretreatment uses biohazard bags before disposal in solid waste trash.

Treatment

Depending on the application, some companies decontaminate with a dose of chlorine dioxide or other deactivator and then dispose of the item in a landfill. This option is more expensive than disposing of an untreated component because it requires extra steps before landfilling. It does, however, allow for the product to be landfilled after use without other cleaning and decontamination steps.

All materials used in the manufacture of disposables contain extractables and leachables. In landfills, where both aerobic (oxidative) and anaerobic (reductive) conditions occur, both of these types of products will be found in the vicinity of deposited waste material. Leachables, by their definition, will be released from plastics under normal landfill conditions. These may also have biohazardous substances associated with them if the waste has not been presterilized, and hence may cause an environmental risk. Extractables may also be released from landfilled plastics over extended periods as aggressive chemical conditions such as acidic, caustic, or even dilute solvent microclimates may be established. Furthermore, high-temperature pretreatment (presterilization) has the potential to convert extractables into leachables. The assessment of the risk associated with extractables and leachables has so far (by definition) been focused on the product and the associated health risks and not on potential environmental risks.

Overall Environmental Impact

While it is generally believed that the environmental impact of conventional stainless steel systems is lower than the disposable systems, the induction of disposable technology carries a negative connotation, particularly in the United States, where 5% of the world's population produces 95% of the world's garbage.

There is a dire need to apprise manufacturers that it is the overall impact on the environment that matters. Several comprehensive studies have confirmed that a fully disposable biopharmaceutical factory can be environmentally advantageous compared with a conventional stainless steel biomanufacturing for the following reasons:

- Disposable systems require 1/10th the water to process an equivalent amount of product. Water is the most precious resource.
- The energy footprint of stainless steel systems far outweighs the energy needed to incinerate plastic waste.
- Despite their long life, the disposal of stainless steel produces a much higher impact since it is not possible to incinerate it.
- Disposable system factories run with much smaller energy requirements and thus add less to the carbon footprint.
- Disposable system factories are less labor intensive, mainly because of elimination of the SIP/CIP and testing for validation.

Summary

One of the greatest impediments in the acceptance of disposable systems has been their image as contributor to the carbon footprint and thus adversely affecting the environment, even in a society that is perhaps the greatest waste producer in the world, the United States. However, most of these perceptions are wrong and due to misinformation from the defenders of the stainless steel industry, which should begin to feel the shudder from these plastic parts. Would there be a time soon when there will be biodegradeable bioreactors? Perhaps. Would there be a time soon when there will be nonleaching bioreactors and other plastic components? Perhaps sooner than later. However, until such changes come about, the manufacturer should realize that a lot of work goes into developing these components, and equipment suppliers are not likely to switch over to different materials from the ones they have had a lot of experience working with. Design and material changes are difficult to make, as it requires a tremendous amount of validation work to ensure that a component would work as it is supposed to every time it is used. Table 12.1 gives an overview of comparisons of various disposal options:

The environmental hazard threat from disposable components is unfounded; more so, when we look at the overall threat by other such risks.

TABLE 12.1

At-a-glance comparison of single-use bioprocess system disposal options

Option	Advantages	Disadvantages
Landfill, untreated	Lowest operating cost, no capital cost	Not an option for hazardous waste; perceived as environmentally unfriendly
Landfill, treated	Inexpensive, no capital cost	Perceived as environmentally unfriendly
Grind, autoclave, and landfill	Generally accepted as safe, reduces landfill volume	Significant capital cost, requires extra handling
Recycling	Environmentally appealing	Impractical for mixed materials
Incinerate	Generally accepted as safe	May be legally restricted and costly
Incinerate with generation of stem or electricity (cogeneration)	Most environmentally benign, some return on investment	May be legally restricted, and presents the highest capital cost
Pyrolysis	Produces usable pure diesel fuel; fuel produced burns more cleanly than that produced from a refinery	New technology—few options available; subpar efficiency

Source: http://www.bpsalliance.org/

Appendix B: Classification of Human Etiologic Agents on the Basis of Hazard

This appendix includes those biological agents known to infect humans as well as selected animal agents that may pose theoretical risks if inoculated into humans. Included are lists of representative genera and species known to be pathogenic; mutated, recombined, and nonpathogenic species and strains are not considered. Noninfectious life-cycle stages of parasites are excluded.

This appendix reflects the current state of knowledge and should be considered a resource document. Included are the more commonly encountered agents and is not meant to be all-inclusive. Information on agent risk assessment may be found in the Agent Summary Statements of the CDC/ NIH publication, Biosafety in Microbiological and Biomedical Laboratories (see Sections V-C, V-D, V-E, and V-F, *Footnotes and References of Sections I through IV.* Further guidance on agents not listed in Appendix B may be obtained through Centers for Disease Control and Prevention, Biosafety Branch, Atlanta, Georgia 30333, Phone: (404) 639-3883, Fax: (404) 639-2294; National Institutes of Health, Division of Safety, Bethesda, Maryland 20892, Phone: (301) 496-1357; National Animal Disease Center, U.S. Department of Agriculture, Ames, Iowa 50010, Phone: (515) 862-8258.

A special committee of the American Society for Microbiology will conduct an annual review of this appendix, and its recommendations for changes will be presented to the Recombinant DNA Advisory Committee as proposed amendments to the NIH Guidelines.

Appendix B-I: Risk Group 1 (RG1) Agents

RG1 agents are not associated with disease in healthy adult humans. Examples of RG1 agents include asporogenic *Bacillus subtilis* or *Bacillus licheniformis* (see Appendix C-IV-A, *Bacillus subtilis or Bacillus licheniformis Host–Vector Systems, Exceptions*); adeno-associated virus (AAV) types 1 through 4; and recombinant AAV constructs, in which the transgene does not encode either a potentially tumorigenic gene product or a toxin molecule and are produced in the absence of a helper virus. A strain of *Escherichia coli* (see Appendix C-II-A, *Escherichia coli K-12 Host–Vector Systems, Exceptions*) is an RG1 agent if it (1) does not possess a complete lipopolysaccharide (i.e., lacks the O antigen); and (2) does not carry any active virulence factor (e.g., toxins) or colonization factors and does not carry any genes encoding these factors.

Those agents not listed in Risk Groups (RGs) 2, 3, and 4 are not automatically or implicitly classified in RG1; a risk assessment must be conducted based on the known and potential properties of the agents and their relationship to agents that are listed.

Appendix B-II: Risk Group 2 (RG2) Agents

RG2 agents are associated with human disease that is rarely serious and for which preventive or therapeutic interventions are often available.

Appendix B-II-A: Risk Group 2 (RG2)— Bacterial Agents Including Chlamydia

Acinetobacter baumannii (formerly *Acinetobacter calcoaceticus*)

Actinobacillus

Actinomyces pyogenes (formerly Corynebacterium pyogenes)

Aeromonas hydrophila

Amycolata autotrophica

Archanobacterium haemolyticum (formerly *Corynebacterium haemolyticum*)

Arizona hinshawii—all serotypes

Bacillus anthracis

Bartonella henselae, B. quintana, B. vinsonii

Bordetella including B. pertussis

Borrelia recurrentis, B. burgdorferi

Burkholderia (formerly Pseudomonas species) except those listed in Appendix B-III-A (RG3))

Campylobacter coli, C. fetus, C. jejuni

Chlamydia psittaci, C. trachomatis, C. pneumoniae

Clostridium botulinum, Cl. chauvoei, Cl. haemolyticum, Cl. histolyticum, Cl. novyi, Cl. septicum, Cl. tetani

Corynebacterium diphtheriae, C. pseudotuberculosis, C. renale

Dermatophilus congolensis

Edwardsiella tarda

Erysipelothrix rhusiopathiae

Escherichia coli—all enteropathogenic, enterotoxigenic, enteroinvasive and strains bearing K1 antigen, including *E. coli* O157:H7

Haemophilus ducreyi, H. influenzae

Helicobacter pylori

Klebsiella—all species except *K. oxytoca* (RG1)

Legionella including *L. pneumophila*

Leptospira interrogans—all serotypes

Listeria

Moraxella

Mycobacterium (except those listed in Appendix B-III-A (RG3)) including M. avium complex, *M. asiaticum, M. bovis* BCG vaccine strain, *M. chelonei, M. fortuitum, M. kansasii, M. leprae, M. malmoense, M. marinum, M. paratuberculosis, M. scrofulaceum, M. simiae, M. szulgai, M. ulcerans, M. xenopi*

Mycoplasma, except *M. mycoides* and *M. agalactiae*, which are restricted animal pathogens

Neisseria gonorrhoeae, N. meningitidis

Nocardia asteroides, N. brasiliensis, N. otitidiscaviarum, N. transvalensis

Rhodococcus equi

Salmonella including *S. arizonae, S. cholerasuis, S. enteritidis, S. gallinarum-pullorum, S. meleagridis, S. paratyphi, A, B, C, S. typhi, S. typhimurium*

Shigella including *S. boydii, S. dysenteriae, type 1, S. flexneri, S. sonnei*

Sphaerophorus necrophorus

Staphylococcus aureus

Streptobacillus moniliformis

Streptococcus including S. *pneumoniae, S. pyogenes*

Treponema pallidum, T. carateum

Vibrio cholerae, V. parahemolyticus, V. vulnificus

Yersinia enterocolitica

Appendix B-II-B: Risk Group 2 (RG2)—Fungal Agents

Blastomyces dermatitidis

Cladosporium bantianum, C. (Xylohypha) trichoides

Cryptococcus neoformans

Dactylaria galopava (Ochroconis gallopavum)

Epidermophyton

Exophiala (Wangiella) dermatitidis

Fonsecaea pedrosoi

Microsporum

Paracoccidioides braziliensis
Penicillium marneffei
Sporothrix schenckii
Trichophyton

Appendix B-II-C: Risk Group 2 (RG2)—Parasitic Agents

Ancylostoma human hookworms including *A. duodenale, A. ceylanicum*

Ascaris including *Ascaris lumbricoides suum*

Babesia including *B. divergens, B. microti*

Brugia filaria worms including *B. malayi, B. timori*

Coccidia

Cryptosporidium including *C. parvum*

Cysticercus cellulosae (hydatid cyst, larva of *T. solium*)

Echinococcus including *E. granulosis, E. multilocularis, E. vogeli*

Entamoeba histolytica

Enterobius

Fasciola including *F. gigantica, F. hepatica*

Giardia including *G. lamblia*

Heterophyes

Hymenolepis including *H. diminuta, H. nana*

Isospora

Leishmania including *L. braziliensis, L. donovani, L. ethiopia, L. major, L. mexicana, L. peruvania, L. tropica*

Loa loa filaria worms

Microsporidium

Naegleria fowleri

Necator human hookworms including *N. americanus*

Onchocerca filaria worms including *O. volvulus*

Plasmodium including simian species, *P. cynomologi, P. falciparum, P. malariae, P. ovale, P. vivax*

Sarcocystis including *S. sui hominis*

Schistosoma including *S. haematobium, S. intercalatum, S. japonicum, S. mansoni, S. mekongi*

Strongyloides including *S. stercoralis*

Taenia solium

Toxocara including *T. canis*

Toxoplasma including *T. gondii*

Trichinella spiralis

Trypanosoma including *T. brucei brucei, T. brucei gambiense, T. brucei rho-desiense, T. cruzi*

Wuchereria bancrofti filaria worms

Appendix B-II-D: Risk Group 2 (RG2)—Viruses

Adenoviruses, human—all types

Alphaviruses (Togaviruses)—Group A Arboviruses

    Eastern equine encephalomyelitis virus

    Venezuelan equine encephalomyelitis vaccine strain TC-83

    Western equine encephalomyelitis virus

Arenaviruses

    Lymphocytic choriomeningitis virus (non-neurotropic strains)

    Tacaribe virus complex

    Other viruses as listed in the reference source (see Section V-C, *Footnotes and References of Sections I through IV*)

Bunyaviruses

    Bunyamwera virus

    Rift Valley fever virus vaccine strain MP-12

    Other viruses as listed in the reference source (see Section V-C, *Footnotes and References of Sections I through IV*)

Caliciviruses

Coronaviruses

Flaviviruses (Togaviruses)—Group B Arboviruses

    Dengue virus serotypes 1, 2, 3, and 4

    Yellow fever virus vaccine strain 17D

    Other viruses as listed in the reference source (see Section V-C, *Footnotes and References of Sections I through IV*)

Hepatitis A, B, C, D, and E viruses

Herpesviruses—except Herpesvirus simiae (Monkey B virus) (see Appendix B-IV-D, Risk Group 4 (RG4)—Viral Agents)

    Cytomegalovirus

    Epstein Barr virus

    Herpes simplex types 1 and 2

    Herpes zoster

    Human herpesvirus types 6 and 7

Orthomyxoviruses

    Influenza viruses types A, B, and C (except those listed in Appendix B-III-D, Risk Group 3 (RG3)—Viruses and Prions)

    Tick-borne orthomyxoviruses

Papovaviruses

    All human papilloma viruses

Paramyxoviruses

    Newcastle disease virus

    Measles virus

    Mumps virus

    Parainfluenza viruses types 1, 2, 3, and 4

    Respiratory syncytial virus

Parvoviruses

    Human parvovirus (B19)

Picornaviruses

    Coxsackie viruses types A and B

    Echoviruses—all types

    Polioviruses—all types, wild and attenuated

    Rhinoviruses—all types

Poxviruses—all types except Monkeypox virus (see Appendix B-III-D, *Risk Group 3 (RG3)—Viruses and Prions*) and restricted poxviruses including Alastrim, Smallpox, and Whitepox (see Section V-L, *Footnotes and References of Sections I through IV*)

Reoviruses—all types including Coltivirus, human Rotavirus, and Orbivirus (Colorado tick fever virus)

Rhabdoviruses

    Rabies virus—all strains

    Vesicular stomatitis virus—laboratory-adapted strains including VSV-Indiana, San Juan, and Glasgow

Togaviruses (see Alphaviruses and Flaviviruses)

Rubivirus (rubella)

Appendix B-III: Risk Group 3 (RG3) Agents

RG3 agents are associated with serious or lethal human disease for which preventive or therapeutic interventions may be available.

Appendix B-III-A: Risk Group 3 (RG3)— Bacterial Agents Including Rickettsia

Bartonella

Brucella including *B. abortus, B. canis, B. suis*

Burkholderia (Pseudomonas) mallei, B. pseudomallei

Coxiella burnetii

Francisella tularensis

Mycobacterium bovis (except BCG strain, see Appendix B-II-A, Risk Group 2 (RG2)—Bacterial Agents Including Chlamydia), *M. tuberculosis*

Pasteurella multocida type B -"buffalo" and other virulent strains

Rickettsia akari, R. australis, R. canada, R. conorii, R. prowazekii, R. rickettsii, R, siberica, R. tsutsugamushi, R. typhi (R. mooseri)

Yersinia pestis

Appendix B-III-B: Risk Group 3 (RG3)—Fungal Agents

Coccidioides immitis (sporulating cultures; contaminated soil)

Histoplasma capsulatum, H. capsulatum var.. duboisii

Appendix B-III-C: Risk Group 3 (RG3)—Parasitic Agents

None

Appendix B-III-D: Risk Group 3 (RG3)—Viruses and Prions

Alphaviruses (Togaviruses)—Group A Arboviruses

Semliki Forest virus

St. Louis encephalitis virus

Venezuelan equine encephalomyelitis virus (except the vaccine strain TC-83; see Appendix B-II-D (RG2))

Other viruses as listed in the reference source (see Section V-C, *Footnotes and References of Sections I through IV*)

Arenaviruses

Flexal

Lymphocytic choriomeningitis virus (LCM) (neurotropic strains)

Bunyaviruses

Hantaviruses including Hantaan virus

Rift Valley fever virus

Flaviviruses (Togaviruses)—Group B Arboviruses

    Japanese encephalitis virus

    Yellow fever virus

    Other viruses as listed in the reference source (see Section V-C, *Footnotes and References of Sections I through IV*)

Orthomyxoviruses

    Influenza viruses 1918-1919 H1N1 (1918 H1N1), human H2N2 (1957-1968), and highly pathogenic avian influenza H5N1 strains within the Goose/Guangdong/96-like H5 lineage (HPAI H5N1)

Poxviruses

    Monkeypox virus

Prions

    Transmissible spongioform encephalopathies (TME) agents (Creutzfeldt-Jacob disease and kuru agents)(see Section V-C, *Footnotes and References of Sections I through IV*, for containment instruction)

Retroviruses

    Human immunodeficiency virus (HIV) types 1 and 2

    Human T cell lymphotropic virus (HTLV) types 1 and 2

    Simian immunodeficiency virus (SIV)

Rhabdoviruses

Vesicular stomatitis virus

Appendix B-IV: Risk Group 4 (RG4) Agents

RG4 agents are likely to cause serious or lethal human disease for which preventive or therapeutic interventions are not usually available.

Appendix B-IV-A: Risk Group 4 (RG4)—Bacterial Agents

None.

Appendix B-IV-B: Risk Group 4 (RG4)—Fungal Agents

None.

Appendix B-IV-C Risk Group 4 (RG4)—Parasitic Agents

None.

Appendix B-IV-D Risk Group 4 (RG4)—Viral Agents

Arenaviruses

    Guanarito virus

    Lassa virus

    Junin virus

    Machupo virus

    Sabia

Bunyaviruses (Nairovirus)

    Crimean-Congo hemorrhagic fever virus

Filoviruses

    Ebola virus

    Marburg virus

Flaviruses (Togaviruses)—Group B Arboviruses

    Tick-borne encephalitis virus complex including Absetterov, Central European encephalitis, Hanzalova, Hypr, Kumlinge, Kyasanur Forest disease, Omsk hemorrhagic fever, and Russian spring-summer encephalitis viruses

Herpesviruses (alpha)

    Herpesvirus simiae (Herpes B or Monkey B virus)

Paramyxoviruses

    Equine morbillivirus

    Hemorrhagic fever agents and viruses as yet undefined

Appendix B-V: Animal Viral Etiologic Agents in Common Use

The following list of animal etiologic agents is appended to the list of human etiologic agents. None of these agents is associated with disease in healthy adult humans; they are commonly used in laboratory experimental work.

A containment level appropriate for RG1 human agents is recommended for their use. For agents that are infectious to human cells, e.g., amphotropic and xenotropic strains of murine leukemia virus, a containment level appropriate for RG2 human agents is recommended.

Baculoviruses

Herpesviruses

Herpesvirus ateles

Herpesvirus saimiri

Marek's disease virus

Murine cytomegalovirus

Papovaviruses

Bovine papilloma virus

Polyoma virus

Shope papilloma virus

Simian virus 40 (SV40)

Retroviruses

Avian leukosis virus

Avian sarcoma virus

Bovine leukemia virus

Feline leukemia virus

Feline sarcoma virus

Gibbon leukemia virus

Mason-Pfizer monkey virus

Mouse mammary tumor virus

Murine leukemia virus

Murine sarcoma virus

Rat leukemia virus

Appendix B-V-1: Murine Retroviral Vectors

Murine retroviral vectors to be used for human transfer experiments (less than 10 L) that contain less than 50% of their respective parental viral genome and that have been demonstrated to be free of detectable replication competent retrovirus can be maintained, handled, and administered, under BL1 containment.

Appendix K: Physical Containment for Large-Scale Uses of Organisms Containing Recombinant DNA Molecules

Appendix K specifies physical containment guidelines for large-scale (greater than 10 liters of culture) research or production involving viable organisms containing recombinant DNA molecules. It shall apply to large-scale research or production activities as specified in Section III-D-6, *Experiments Involving More than 10 Liters of Culture*. It is important to note that this appendix addresses only the biological hazard associated with organisms containing recombinant DNA. Other hazards accompanying the large-scale cultivation of such organisms (e.g., toxic properties of products; physical, mechanical, and chemical aspects of downstream processing) are not addressed and shall be considered separately, albeit in conjunction with this appendix.

All provisions shall apply to large-scale research or production activities with the following modifications: (i) Appendix K shall supersede Appendix G, *Physical Containment*, when quantities in excess of 10 liters of culture are involved in research or production. Appendix K-II applies to Good Large-Scale Practice; (ii) the institution shall appoint a Biological Safety Officer if it engages in large-scale research or production activities involving viable organisms containing recombinant DNA molecules. The duties of the Biological Safety Officer shall include those specified in Section IV-B-3, *Biological Safety Officer*; (iii) the institution shall establish and maintain a health surveillance program for personnel engaged in large-scale research or production activities involving viable organisms containing recombinant DNA molecules which require Biosafety Level (BL) 3 containment at the laboratory scale. The program shall include: preassignment and periodic physical and medical examinations; collection, maintenance, and analysis of serum specimens for monitoring serologic changes that may result from the employee's work experience; and provisions for the investigation of any serious, unusual, or extended illnesses of employees to determine possible occupational origin.

Appendix K-I: Selection of Physical Containment Levels

The selection of the physical containment level required for recombinant DNA research or production involving more than 10 liters of culture is based on the containment guidelines established in Section III, *Experiments Covered by the NIH Guidelines*. For purposes of large-scale research or production, four physical containment levels are established. The four levels set containment conditions at those appropriate for the degree of hazard to health or the environment posed by the organism, judged by experience with similar organisms unmodified by recombinant DNA techniques and consistent with Good Large-Scale Practice. The four biosafety levels of large-scale physical containment are referred to as Good Large-Scale Practice, BL1-Large Scale, BL2-Large Scale, and BL3-Large Scale. Good Large-Scale Practice is recommended for large-scale research or production involving viable, nonpathogenic, and non-toxigenic recombinant strains derived from host organisms that have an extended history of safe large-scale use. Good Large-Scale Practice is recommended for organisms such as those included in Appendix C, *Exemptions under Section III-F-6*, which have built-in environmental limitations that permit optimum growth in the large-scale setting but limited survival without adverse consequences in the environment. BL1-Large Scale is recommended for large-scale research or production of viable organisms containing recombinant DNA molecules that require BL1 containment at the laboratory scale and that do not qualify for Good Large-Scale Practice. BL2-Large Scale is recommended for large-scale research or production of viable organisms containing recombinant DNA molecules that require BL2 containment at the laboratory scale. BL3-Large Scale is recommended for large-scale research or production of viable organisms containing recombinant DNA molecules that require BL3 containment at the laboratory scale. No provisions are made for large-scale research or production of viable organisms containing recombinant DNA molecules that require BL4 containment at the laboratory scale. If necessary, these requirements will be established by NIH on an individual basis.

Appendix K–II: Good Large-Scale Practice (GLSP)

Appendix K-II-A. Institutional codes of practice shall be formulated and implemented to assure adequate control of health and safety matters.

Appendix K-II-B. Written instructions and training of personnel shall be provided to assure that cultures of viable organisms containing recombinant DNA molecules are handled prudently and that the workplace is kept clean and orderly.

Appendix K-II-C. In the interest of good personal hygiene, facilities (e.g., hand washing sink, shower, changing room) and protective clothing (e.g., uniforms, laboratory coats) shall be provided that are appropriate for the risk of exposure to viable organisms containing recombinant DNA molecules. Eating, drinking, smoking, applying cosmetics, and mouth pipetting shall be prohibited in the work area.

Appendix K-II-D. Cultures of viable organisms containing recombinant DNA molecules shall be handled in facilities intended to safeguard health during work with microorganisms that do not require containment.

Appendix K-II-E. Discharges containing viable recombinant organisms shall be handled in accordance with applicable governmental environmental regulations.

Appendix K-II-F. Addition of materials to a system, sample collection, transfer of culture fluids within/between systems, and processing of culture fluids shall be conducted in a manner that maintains employees' exposure to viable organisms containing recombinant DNA molecules at a level that does not adversely affect the health and safety of employees.

Appendix K-II-G. The facility's emergency response plan shall include provisions for handling spills.

Appendix K-III: Biosafety Level 1 (BL1)—Large Scale

Appendix K-III-A. Spills and accidents that result in overt exposures to organisms containing recombinant DNA molecules are immediately reported to the Laboratory Director. Medical evaluation, surveillance, and treatment are provided as appropriate, and written records are maintained.

Appendix K-III-B. Cultures of viable organisms containing recombinant DNA molecules shall be handled in a closed system (e.g., closed vessel used for the propagation and growth of cultures) or other primary containment equipment (e.g., biological safety cabinet containing a centrifuge used to process culture fluids), which is designed to reduce the potential for escape of viable organisms. Volumes less than 10 liters may be handled outside of a closed system or other primary containment equipment provided all physical containment requirements specified in Appendix G-II-A, *Physical Containment Levels—Biosafety Level 1*, are met.

Appendix K-III-C. Culture fluids (except as allowed in Appendix K-III-D) shall not be removed from a closed system or other primary containment equipment unless the viable organisms containing recombinant DNA molecules have been inactivated by a validated inactivation procedure. A validated inactivation procedure is one that has been demonstrated to be effective using the organism that will serve as the host for propagating the recombinant DNA molecules. Culture fluids that contain viable organisms or viral vectors intended as the final product may be removed from the primary containment equipment by way of closed systems for sample analysis, further processing, or final fill.

Appendix K-III-D. Sample collection from a closed system, the addition of materials to a closed system, and the transfer of culture fluids from one closed system to another shall be conducted in a manner that minimizes the release of aerosols or contamination of exposed surfaces.

Appendix K-III-E. Exhaust gases removed from a closed system or other primary containment equipment shall be treated by filters which have efficiencies equivalent to high efficiency particulate air/HEPA filters or by other equivalent procedures (e.g., incineration) to minimize the release of viable organisms containing recombinant DNA molecules to the environment.

Appendix K-III-F. A closed system or other primary containment equipment that has contained viable organisms containing recombinant DNA molecules shall not be opened for maintenance or other purposes unless it has been sterilized by a validated sterilization procedure except when the culture fluids contain viable organisms or vectors intended as the final product as described in Appendix K-III-C above. A validated sterilization procedure is one that has been demonstrated to be effective using the organism that will serve as the host for propagating the recombinant DNA molecules.

Appendix K-III-G. Emergency plans required by Sections IV-B-2-b-(6), *Institutional Biosafety Committee*, and IV-B-3-c-(3), *Biological Safety Officer*, shall include methods and procedures for handling large losses of culture on an emergency basis.

Appendix K-IV: Biosafety Level 2 (BL2)—Large Scale

Appendix K-IV-A. Spills and accidents that result in overt exposures to organisms containing recombinant DNA molecules are immediately reported to the Biological Safety Officer, Institutional Biosafety Committee, NIH/OBA, and other appropriate authorities (if applicable). Reports to NIH/OBA shall be sent to the Office of Biotechnology Activities, National Institutes of Health, 6705 Rockledge Drive, Suite 750, MSC 7985, Bethesda, MD 20892-7985 (20817 for non-USPS mail), 301-496-9838, 301-496-9839 (fax). Medical evaluation, surveillance, and treatment are provided as appropriate, and written records are maintained.

Appendix K-IV-B. Cultures of viable organisms containing recombinant DNA molecules shall be handled in a closed system (e.g., closed vessel used for the propagation and growth of cultures) or other primary containment equipment (e.g., Class III biological safety cabinet containing a centrifuge used to process culture fluids), which is designed to prevent the escape of viable organisms. Volumes less than 10 liters may be handled outside of a closed system or other primary containment equipment provided all physical containment requirements specified in Appendix G-II-B, *Physical Containment Levels--Biosafety Level 2*, are met.

Appendix K-IV-C. Culture fluids (except as allowed in Appendix K-IV-D) shall not be removed from a closed system or other primary containment equipment unless the viable organisms containing recombinant DNA molecules have been inactivated by a validated inactivation procedure. A validated inactivation procedure is one that has been demonstrated to be effective using the organism that will serve as the host for propagating the recombinant DNA molecules. Culture fluids that contain viable organisms or viral vectors intended as the final product may be removed from the primary containment equipment by way of closed systems for sample analysis, further processing, or final fill.

Appendix K-IV-D. Sample collection from a closed system, the addition of materials to a closed system, and the transfer of cultures fluids from one closed system to another shall be conducted in a manner that prevents the release of aerosols or contamination of exposed surfaces.

Appendix K-IV-E. Exhaust gases removed from a closed system or other primary containment equipment shall be treated by filters that have efficiencies equivalent to high efficiency particulate air/HEPA filters or by other equivalent procedures (e.g., incineration) to prevent the release of viable organisms containing recombinant DNA molecules to the environment.

Appendix K-IV-F. A closed system or other primary containment equipment that has contained viable organisms containing recombinant DNA molecules shall not be opened for maintenance or other purposes unless it has been sterilized by a validated sterilization procedure except when the culture fluids contain viable organisms or vectors intended as final product as described in Appendix K-IV-C above. A validated sterilization procedure is one that has been demonstrated to be effective using the organisms that will serve as the host for propagating the recombinant DNA molecules.

Appendix K-IV-G. Rotating seals and other mechanical devices directly associated with a closed system used for the propagation and growth of viable organisms containing recombinant DNA molecules shall be designed to prevent leakage or shall be fully enclosed in ventilated housings that are exhausted through filters that have efficiencies equivalent to high efficiency particulate air/HEPA filters or through other equivalent treatment devices.

Appendix K-IV-H. A closed system used for the propagation and growth of viable organisms containing recombinant DNA molecules and other primary containment equipment used to contain operations involving viable organisms containing sensing devices that monitor the integrity of containment during operations.

Appendix K-IV-I. A closed system used for the propagation and growth of viable organisms containing the recombinant DNA molecules shall be tested for integrity of the containment features using the organism that will serve as the host for propagating recombinant DNA molecules. Testing shall be accomplished prior to the introduction of viable organisms containing recombinant DNA molecules and following modification or replacement of essential containment features. Procedures and methods used in the testing shall be appropriate for the equipment design and for recovery and demonstration of the test organism. Records of tests and results shall be maintained on file.

Appendix K-IV-J. A closed system used for the propagation and growth of viable organisms containing recombinant DNA molecules shall be permanently identified. This identification shall be used in all records reflecting testing, operation, and maintenance and in all documentation relating to use of this equipment for research or production activities involving viable organisms containing recombinant DNA molecules.

Appendix K-IV-K. The universal biosafety sign shall be posted on each closed system and primary containment equipment when used to contain viable organisms containing recombinant DNA molecules.

Appendix K-IV-L. Emergency plans required by Sections IV-B-2-b-(6), *Institutional Biosafety Committee*, and IV-B-3-c-(3), *Biological Safety Officer*, shall include methods and procedures for handling large losses of culture on an emergency basis.

Appendix K-V: Biosafety Level 3 (BL3)—Large Scale

Appendix K-V-A. Spills and accidents that result in overt exposures to organisms containing recombinant DNA molecules are immediately reported to the Biological Safety Officer, Institutional Biosafety Committee, NIH/OBA, and other appropriate authorities (if applicable). Reports to NIH/OBA shall be sent to the Office of Biotechnology Activities, National Institutes of Health, 6705 Rockledge Drive, Suite 750, MSC 7985, Bethesda, MD 20892-7985 (20817 for non-USPS mail), 301-496-9838, 301-496-9839 (fax). Medical evaluation, surveillance, and treatment are provided as appropriate and written records are maintained.

Appendix K-V-B. Cultures of viable organisms containing recombinant DNA molecules shall be handled in a closed system (e.g., closed vessels used for the propagation and growth of cultures) or other primary containment equipment (e.g., Class III biological safety cabinet containing a centrifuge used to process culture fluids), which is designed to prevent the escape of viable organisms. Volumes less than 10 liters may be handled outside of a closed system provided all physical containment requirements specified in Appendix G-II-C, *Physical Containment Levels—Biosafety Level 3*, are met.

Appendix K-V-C. Culture fluids (except as allowed in Appendix K-V-D) shall not be removed from a closed system or other primary containment equipment unless the viable organisms containing recombinant DNA molecules have been inactivated by a validated inactivation procedure. A validated inactivation procedure is one that has been demonstrated to be effective using the organisms that will serve as the host for propagating the recombinant DNA molecules. Culture fluids that contain viable organisms or viral vectors intended as the final product may be removed from the primary containment equipment by way of closed systems for sample analysis, further processing, or final fill.

Appendix K-V-D. Sample collection from a closed system, the addition of materials to a closed system, and the transfer of culture fluids from one closed system to another shall be conducted in a manner that prevents the release of aerosols or contamination of exposed surfaces.

Appendix K-V-E. Exhaust gases removed from a closed system or other primary containment equipment shall be treated by filters that have efficiencies equivalent to high efficiency particulate air/HEPA filters or by other equivalent procedures (e.g., incineration) to prevent the release of viable organisms containing recombinant DNA molecules to the environment.

Appendix K-V-F. A closed system or other primary containment equipment that has contained viable organisms containing recombinant DNA molecules shall not be opened for maintenance or other purposes unless it has been sterilized by a validated sterilization procedure except when the culture fluids contain viable organisms or vectors intended as the final product as described in Appendix K-V-C above. A validated sterilization procedure is one that has been demonstrated to be effective using the organisms that will serve as the host for propagating the recombinant DNA molecules.

Appendix K-V-G. A closed system used for the propagation and growth of viable organisms containing recombinant DNA molecules shall be operated so that the space above the culture level will be maintained at a pressure as low as possible, consistent with equipment design, in order to maintain the integrity of containment features.

Appendix K-V-H. Rotating seals and other mechanical devices directly associated with a closed system used to contain viable organisms containing recombinant DNA molecules shall be designed to prevent leakage or shall be fully enclosed in ventilated housings that are exhausted through filters that have efficiencies equivalent to high efficiency particulate air/HEPA filters or through other equivalent treatment devices.

Appendix K-V-I. A closed system used for the propagation and growth of viable organisms containing recombinant DNA molecules and other primary containment equipment used to contain operations involving viable organisms containing recombinant DNA molecules shall include monitoring or sensing devices that monitor the integrity of containment during operations.

Appendix K-V-J. A closed system used for the propagation and growth of viable organisms containing recombinant DNA molecules shall be tested for integrity of the containment features using the organisms that will serve as the host for propagating the recombinant DNA molecules. Testing shall be accomplished prior to the introduction of viable organisms containing recombinant DNA molecules and following modification or replacement of essential containment features. Procedures and methods used in the testing shall be appropriate for the equipment design and for recovery and demonstration of the test organism. Records of tests and results shall be maintained on file.

Appendix K-V-K. A closed system used for the propagation and growth of viable organisms containing recombinant DNA molecules shall be permanently identified. This identification shall be used in all records reflecting testing, operation, maintenance, and use of this equipment for research production activities involving viable organisms containing recombinant DNA molecules.

Appendix K-V-L. The universal biosafety sign shall be posted on each closed system and primary containment equipment when used to contain viable organisms containing recombinant DNA molecules.

Appendix K-V-M. Emergency plans required by Sections IV-B-2-b-(6), *Institutional Biosafety Committee*, and IV-B-3-c-(3), *Biological Safety Officer*, shall include methods and procedures for handling large losses of culture on an emergency basis.

Appendix K-V-N. Closed systems and other primary containment equipment used in handling cultures of viable organisms containing recombinant DNA molecules shall be located within a controlled area that meets the following requirements:

Appendix K-V-N-1. The controlled area shall have a separate entry area. The entry area shall be a double-doored space such as an air lock, anteroom, or change room that separates the controlled area from the balance of the facility.

Appendix K-V-N-2. The surfaces of walls, ceilings, and floors in the controlled area shall be such as to permit ready cleaning and decontamination.

Appendix K-V-N-3. Penetrations into the controlled area shall be sealed to permit liquid or vapor phase space decontamination.

Appendix K-V-N-4. All utilities and service or process piping and wiring entering the controlled area shall be protected against contamination.

Appendix K-V-N-5. Hand-washing facilities equipped with foot, elbow, or automatically operated valves shall be located at each major work area and near each primary exit.

Appendix K-V-N-6. A shower facility shall be provided. This facility shall be located in close proximity to the controlled area.

Appendix K-V-N-7. The controlled area shall be designed to preclude release of culture fluids outside the controlled area in the event of an accidental spill or release from the closed systems or other primary containment equipment.

Appendix K-V-N-8. The controlled area shall have a ventilation system that is capable of controlling air movement. The movement of air shall be from areas of lower contamination potential to areas of higher contamination potential. If the ventilation system provides positive pressure supply air, the system shall operate in a manner that prevents the reversal of the direction of air movement or shall be equipped with an alarm that would be actuated in the event that reversal in the direction of air movement were to occur. The exhaust air from the controlled area shall not be recirculated to other areas of the facility. The exhaust air from the controlled area may not be discharged to the outdoors without being high efficiency particulate air/HEPA filtered, subjected to thermal oxidation, or otherwise treated to prevent the release of viable organisms.

Appendix K-V-O. The following personnel and operational practices shall be required:

Appendix K-V-O-1. Personnel entry into the controlled area shall be through the entry area specified in Appendix K-V-N-1.

Appendix K-V-O-2. Persons entering the controlled area shall exchange or cover their personal clothing with work garments such as jumpsuits, laboratory coats, pants and shirts, head cover, and shoes or shoe covers. On exit from the controlled area the work clothing may be stored in a locker separate from that used for personal clothing or discarded for laundering. Clothing shall be decontaminated before laundering.

Appendix K-V-O-3. Entry into the controlled area during periods when work is in progress shall be restricted to those persons required to meet program or support needs. Prior to entry, all persons shall be informed of the operating practices, emergency procedures, and the nature of the work conducted.

Appendix K-V-O-4. Persons under 18 years of age shall not be permitted to enter the controlled area.

Appendix K-V-O-5. The universal biosafety sign shall be posted on entry doors to the controlled area and all internal doors when any work involving the organism is in progress. This includes periods when decontamination procedures are in progress. The sign posted on the entry doors to the controlled area shall include a statement of agents in use and personnel authorized to enter the controlled area.

Appendix K-V-O-6. The controlled area shall be kept neat and clean.

Appendix K-V-O-7. Eating, drinking, smoking, and storage of food are prohibited in the controlled area.

Appendix K-V-O-8. Animals and plants shall be excluded from the controlled area.

Appendix K-V-O-9. An effective insect and rodent control program shall be maintained.

Appendix K-V-O-10. Access doors to the controlled area shall be kept closed, except as necessary for access, while work is in progress. Serve doors leading directly outdoors shall be sealed and locked while work is in progress.

Appendix K-V-O-11. Persons shall wash their hands when exiting the controlled area.

Appendix K-V-O-12. Persons working in the controlled area shall be trained in emergency procedures.

Appendix K-V-O-13. Equipment and materials required for the management of accidents involving viable organisms containing recombinant DNA molecules shall be available in the controlled area.

Appendix K-V-O-14. The controlled area shall be decontaminated in accordance with established procedures following spills or other accidental release of viable organisms containing recombinant DNA molecules.

Appendix K-VI: Footnotes of Appendix K

Appendix K-VI-A. Table K.1 is derived from the text in Appendices G (*Physical Containment*) and K and is not to be used in lieu of Appendices G and K.

TABLE K.1

Comparison of Good Large-Scale Practice (GLSP) and Biosafety Level (BL)—
Large-Scale (LS) Practice

CRITERION [See Appendix K-VI-B, Footnotes of Appendix K]		GLSP	BL1-LS	BL2-LS	BL3-LS
1.	Formulate and implement institutional codes of practice for safety of personnel and adequate control of hygiene and safety measures.	K-II-A	G-I		
2.	Provide adequate written instructions and training of personnel to keep workplace clean and tidy and to keep exposure to biological, chemical, or physical agents at a level that does not adversely affect health and safety of employees.	K-II-B	G-I		
3.	Provide changing and hand-washing facilities as well as protective clothing, appropriate to the risk, to be worn during work.	K-II-C	G-II-A-1-h	G-II-B-2-f	G-II-C-2-i
4.	Prohibit eating, drinking, smoking, mouth pipetting, and applying cosmetics in the work place.	K-II-C	G-II-A-1-d G-II-A-1-e	G-II-B-1-d G-II-B-1-e	G-II-C-1-c G-II-C-1-d
5.	Internal accident reporting.	K-II-G	K-III-A	K-IV-A	K-V-A
6.	Medical surveillance.	NR	NR		
7.	Viable organisms should be handled in a system that physically separates the process from the external environment (closed system or other primary containment).	NR	K-III-B	K-IV-B	K-V-B
8.	Culture fluids not removed from a system until organisms are inactivated.	NR	K-III-C	K-IV-C	K-V-C
9.	Inactivation of waste solutions and materials with respect to their biohazard potential.	K-II-E	K-III-C	K-IV-C	K-V-C

Continued

TABLE K.1 (Continued)

Comparison of Good Large-Scale Practice (GLSP) and Biosafety Level (BL)— Large-Scale (LS) Practice

CRITERION [See Appendix K-VI-B, Footnotes of Appendix K]	GLSP	BL1-LS	BL2-LS	BL3-LS
10. Control of aerosols by engineering or procedural controls to prevent or minimize release of organisms during sampling from a system, addition of materials to a system, transfer of cultivated cells, and removal of material, products, and effluent from a system.	Minimize *Procedure* K-II-F	Minimize *Engineer* K-III-B K-III-D	Prevent *Engineer* K-IV-B K-IV-D	Prevent *Engineer* K-V-B K-V-D
11. Treatment of exhaust gases from a closed system to minimize or prevent release of viable organisms.	NR	Minimize K-III-E	Prevent K-IV-E	Prevent K-V-E
12. Closed system that has contained viable organisms not to be opened until sterilized by a validated procedure.	NR	K-III-F	K-IV-F	K-V-F
13. Closed system to be maintained at as low a pressure as possible to maintain the integrity of containment features.	NR	NR	NR	K-V-G
14. Rotating seals and other penetrations into closed system designed to prevent or minimize leakage.	NR	NR	Prevent K-IV-G	Prevent K-V-H
15. Closed system shall incorporate monitoring or sensing devices to monitor the integrity of containment.	NR	NR	K-IV-H	K-V-I
16. Validated integrity testing of closed containment system.	NR	NR	K-IV-I	K-V-J
17. Closed system to be permanently identified for record keeping purposes.	NR	NR	K-IV-J	K-V-K
18. Universal biosafety sign to be posted on each closed system.	NR	NR	K-IV-K	K-V-L
19. Emergency plans required for handling large losses of cultures.	K-II-G	K-III-G	K-IV-L	K-V-M

Continued

TABLE K.1 (Continued)

Comparison of Good Large-Scale Practice (GLSP) and Biosafety Level (BL)—
Large-Scale (LS) Practice

CRITERION [See Appendix K-VI-B, Footnotes of Appendix K]	GLSP	BL1-LS	BL2-LS	BL3-LS
20. Access to the workplace.	NR	G-II-A-1-a	G-II-B-1-a	K-V-N
21. Requirements for controlled access area.	NR	NR	NR	K-V-N&O

Note: NR = not required. The criteria in this grid address only the biological hazards associated with organisms containing recombinant DNA. Other hazards accompanying the large-scale cultivation of such organisms (e.g., toxic properties of products; physical, mechanical, and chemical aspects of downstream processing) are not addressed and shall be considered separately, albeit in conjunction with this grid.

Appendix K-VI-B. The criteria in this grid address only the biological hazards associated with organisms containing recombinant DNA. Other hazards accompanying the large-scale cultivation of such organisms (e.g., toxic properties of products; physical, mechanical, and chemical aspects of downstream processing) are not addressed and shall be considered separately, albeit in conjunction with this grid.

Appendix K-VII: Definitions to Accompany Containment Grid and Appendix K

Appendix K-VII-A. Accidental Release. An accidental release is the unintentional discharge of a microbiological agent (i.e., microorganism or virus) or eukaryotic cell due to a failure in the containment system.

Appendix K-VII-B. Biological Barrier. A biological barrier is an impediment (naturally occurring or introduced) to the infectivity and/or survival of a microbiological agent or eukaryotic cell once it has been released into the environment.

Appendix K-VII-C. Closed System. A closed system is one in which by its design and proper operation, prevents release of a microbiological agent or eukaryotic cell contained therein.

Appendix K-VII-D. Containment. Containment is the confinement of a microbiological agent or eukaryotic cell that is being cultured, stored, manipulated, transported, or destroyed in order to prevent or limit its contact with people and/or the environment. Methods used to achieve this include physical and biological barriers and inactivation using physical or chemical means.

Appendix K-VII-E. De minimis Release. De minimis release is the release of: (i) viable microbiological agents or eukaryotic cells that does not result in the establishment of disease in healthy people, plants, or animals; or (ii) in uncontrolled proliferation of any microbiological agents or eukaryotic cells.

Appendix K-VII-F. Disinfection. Disinfection is a process by which viable microbiological agents or eukaryotic cells are reduced to a level unlikely to produce disease in healthy people, plants, or animals.

Appendix K-VII-G. Good Large-Scale Practice Organism. For an organism to qualify for Good Large-Scale Practice consideration, it must meet the following criteria [Reference: Organization for Economic Cooperation and Development, *Recombinant DNA Safety Considerations*, 1987, p. 34–35]: (i) the host organism should be nonpathogenic, should not contain adventitious agents and should have an extended history of safe large-scale use or have built-in environmental limitations that permit optimum growth in the large-scale setting but limited survival without adverse consequences in the environment; (ii) the recombinant DNA-engineered organism should be non-pathogenic, should be as safe in the large-scale setting as the host organism, and without adverse consequences in the environment; and (iii) the vector/insert should be well characterized and free from known harmful sequences; should be limited in size as much as possible to the DNA required to perform the intended function; should not increase the stability of the construct in the environment unless that is a requirement of the intended function; should be poorly mobilizable; and should not transfer any resistance markers to microorganisms unknown to acquire them naturally if such an acquisition could compromise the use of a drug to control disease agents in human or veterinary medicine or agriculture.

Appendix K-VII-H. Inactivation. Inactivation is any process that destroys the ability of a specific microbiological agent or eukaryotic cell to self-replicate.

Appendix K-VII-I. Incidental Release. An incidental release is the discharge of a microbiological agent or eukaryotic cell from a containment system that is expected when the system is appropriately designed and properly operated and maintained.

Appendix K-VII-J. Minimization. Minimization is the design and operation of containment systems in order that any incidental release is a de minimis release.

Appendix K-VII-K. Pathogen. A pathogen is any microbiological agent or eukaryotic cell containing sufficient genetic information, which upon expression of such information, is capable of producing disease in healthy people, plants, or animals.

Appendix K-VII-L. Physical Barrier. A physical barrier is considered any equipment, facilities, or devices (e.g., fermentors, factories, filters, thermal oxidizers), which are designed to achieve containment.

Appendix K-VII-M. Release. Release is the discharge of a microbiological agent or eukaryotic cell from a containment system. Discharges can be incidental or accidental. Incidental releases are de minimis in nature; accidental releases may be de minimis in nature.

13

Epilogue

When someone says, 'It's not the money, it's the principle of the thing,' it's the money.

Kin Hubbard

A patent is a set of exclusive rights granted by a state (national government) to an inventor or an assignee thereof for a limited period of time in exchange for the public disclosure of an invention so that humanity can benefit from it. Humanity would not have fared well had it not been for its inventive nature. From the wheel to wheelbarrow, from microscopy to recombinant engineering, inventions have continued to surprise people of their own ability. The most difficult challenge an invention faces is not in demonstrating its usefulness; it is in making others believe that it indeed is good for them. Inventions are viewed as invasive, likely to disrupt a comfort zone, and of little value in the minds of those who are rigid-minded, as most of humanity is. The great inventor Benjamin Franklin was bounced around by his friends as an eccentric; the Wrights had great difficulty selling their idea; and Gregor Mendel's discovery of basic genetic tendencies in his research on peas was overlooked for 50 years.

Long ago, Plato raised a concern in his *Phaedrus* that is familiar in our era: new technology will undermine traditional literacy. Plato (quoting Socrates) expressed the fear that the emerging technology of writing would destroy the rich oral literature that was central to his culture. Writing would reduce the need for memory and attentive listening; it would give learners the appearance of wisdom by aiding rapid recall of information and facts without requiring internalization of such wisdom. This sort of "superficial" learner would inevitably be less literate. It turned out Plato was right only in part: although writing did change the meaning of literacy, it enabled incredible advancements in knowledge.

The disposable technology of today was only a tinkertoy a couple of decades ago. In an industry flushed with cash and used in mega projects, a flexible bag for manufacturing drugs just did not fit the picture well. Still today, neither the FDA nor the EMEA have any product that was manufactured in a disposable bioreactor. Until such time that this taboo is shattered, there shall remain many critics of the disposable bioprocessing systems.

The industry needs to realize that adopting disposal processing systems would improve their profit margins: there is money to be made by switching. Only then would disposable technology finds its rightful place, and all other criticisms, ranging from accusations that it is environment unfriendly to it being impractical, would disappear.

So, for the sake of posterity, here are some predictions for the future of disposable bioprocessing.

Large Scale

The industry used to 10,000 to 100,000 L multistory fermenters is highly suspicious of the scalability of manufacturing using smaller-size disposable bioreactors. The criticism is well-founded, except there is really no need to have those behemoth reactors in the first place. The cell lines of yore produced subgram yield; today it is hovering around 10 g/L, which reduces the size required to about 5% of what it was in the past. A 5 g/L monoclonal cell line would easily produce 100 kg of material (enough for a generic launch of the product), with 20,000 L per year or ten batches from a 2,000 L bioreactor: a size that is currently available for disposable systems.

In the future, manufacturers would not insist on larger bioreactors but better cell lines and connecting the bioreactors in a daisy chain to produce CFR 21-compliant variable size batches. See www.mayabio.com.

Integrity

The integrity of large-size bags (1 to 3,000 L) is often questioned; these sizes are needed to hold media, buffer solutions, and other intermediate products. It is true that flexible bags do not have the strength to hold such large weights, but they can always be supported by a cage around them to distribute the weight on the seams and body of the bags. In the future, one will see engineered dimensions of bags that would eliminate any limitation on the size of fluids that could be held inside a flexible bag.

Flexibility

Future manufacturing will need to be flexible to handle multiple products, different production volumes for each product, and rapid changes in market demand at lower costs. However, many of today's facilities are built to supply blockbuster-like products with high volume and steady demand. The fixed configuration is usually product specific or process specific. Introducing new products into such facilities often requires modifications that can be expensive with long lead times. It is especially challenging to scale production up or down to market demand as both directions incur financial consequences, either with significant capital investment or facility charges for idle capacity. Modular disposable systems would solve all of these constraints.

There is somewhere between 1.1 million and 1.2 million liters of excess capacity in the stainless steel cell culture and fermentation industry; the majority of which is represented by reactors exceeding 2,000 L. With increasing yields, smaller budgets for development, and increase in the diversity of development projects, the future of bioprocessing rests on the adoption of smaller-size bioreactors that would be scalable but also flexible in their size, capabilities, and regulatory compliance. The cost of operating a 10,000 L bioreactor is well established; unfortunately, it is not cost-effective to operate it at a smaller than minimum level: the gap will be filled by disposable bioreactors that can be readied almost instantly and allow variable size production volumes. Flexibility in the future processing industry would expedite the speed of drug delivery, allowing many companies to make it to the market first and allowing the expansion of biotechnology in such areas as cell therapy, stem cell applications, and organ growth simulations.

Universal Use

Stainless steel bioreactors and fermenters have long served the industry well as different cell lines and organisms were inducted as the workhorses. There is no doubt about the versatility of this equipment. Disposable bioreactors first appeared as flexible bags and in a 2D design that was ill-suited to achieve the high K_La values to grow bacteria and yeast; they served well for growing animal cells such as Chinese hamster ovary (CHO). The manufacturing industry did not embrace this well: why create separate systems for bacterial culture growth and mammalian cell cultures if there were a reactor available to growth bacteria? This could be easily modified to grow animal cells. Compounded by the limited size of the disposable bioreactors available initially, the idea of adopting them in the mainstream was lost. To address this shortcoming, the major equipment manufacturers began developing 3D bioreactors wherein they could install similar mixing and aeration systems as used in the stainless steel systems. Today, two companies, Xcellerex and Thermo Scientific Hyclone, offer 2,000 L bioreactors for bacterial fermentation. These are 3D stainless steel shells lined with disposable liners embedded with mixers and aerators, just like the traditional stainless steel bioreactors. The cost of ownership is perhaps higher than the stainless steel system. While these options do offer many of the advantages of disposable systems, the cost advantage can only be had if the industry starts using 2D bioreactors instead, which are much cheaper to construct, require no shell structure, and can be operated horizontally. To overcome these technological barriers, MayaBio developed stationary 2D bioreactors (www.mayabio.com): by removing the movement of the bag (as it is required by GE's Wave), the problems of integrity and size were eliminated. A horizontally lying 2D bag can be used as long as needed; it is heated from underneath the table, and the liquid inside the bag is moved by a patented flapper system that pushes down on the bag to create a wave inside the bag. A slight inclination of the bag provides both kinetic and potential energy to the medium, and K_La values into 100+ are readily obtained by aerating the media inside the bag using a proprietary ceramic tube sparger. These 2D disposable bags can be used to grow every cell and organism, and at a confluence and optical density (OD) that is traditionally seen in stainless steel reactors, all at a fraction of the cost of stainless steel reactors and 3D disposable bioreactors. In the future, most bioreactors will be of the 2D type at least in the fields of drug development, contract manufacturing organizations (CMOs), contact research organizations (CROs), research institutions, and small companies.

Scale-Up

While clinical supplies can be prepared on a smaller bioreactor, the manufacturer wants to make sure that larger volumes will be available fully scaled-up; this was not possible with the best-selling Wave Bioreactor as it was limited to a size of 500 L and there were no options to go to higher volumes. Since the Wave Bioreactor requires rocking the bag, the volume of 500 L was just about what the bag would take as stress; also moving half a ton of liquid up and down required some heavy engineering. What Wave could have done was to promote the use of multiple bioreactors in a chain to obtain higher batch sizes, an idea that was recently innovated by MayaBio that used the Wave Bioreactors to transform them into producing a high growth of bacterial cultures. Since manufacturers must meet the CFR 21 definition of a batch as being a homogenous system, it is possible to circulate the media among several bags to qualify them as a single batch to reduce the cost of testing several batches. In the future, this method will be widely used to reduce the validation requirements for different size batches, and entails only conducting mixing validation studies.

Cost

The current market of major disposable bioprocessing equipment is controlled by a few giants: GE, Pall, Sartorius-Stedim, EMD Millipore, Thermo Fisher Scientific, Xcellerex, and Saint-Gobain. A large number of small, specialized producers of equipment are filling the need of small-to-medium-sized bioreactors. It is not surprising that many of the leaders are emulating each other to stay competitive in terms of the choices available. However, the cost of these integrated systems offered by GE, EMD Millipore, and Sartorius-Stedim has become prohibitive for use on a small scale. Once the equipment manufacturers realize that there is a large market to serve with less expensive systems, the prices will fall. However, there will always be the higher-end products for Big Pharma, which loves to spend big bucks. For the rest of the world, there will be more reasonable choices.

Out of Steam

For over a hundred years, the pharmaceutical and biotechnology industries have had a stereotyped design: every company would have a boiler, a deionization unit, a distillation unit, a storage tank for distilled water, and a stainless steel loop running at 80°C to supply WFI to the point of use where it would be cooled down to make it useable. There would also be huge autoclaves with hundreds of square meters of shelf space fed by clean steam generated from WFI. All of this is about to change in the bioprocessing industry. Today, it is possible to design a facility without SIP/CIP systems and even without autoclaves. This method of going bare will take a while for the industries to absorb, but this is indeed the future of bioprocessing. The result: major savings, conservation of energy, and protection of the environment by conserving water.

Validation

The validation of disposable systems has not been put to the real test yet where a regulatory agency would provide approval of a product manufactured using disposable systems. There is also an unfulfilled need for protocols to run process analytic technology (PAT) on disposable systems. While many new devices have appeared in the market to monitor noninvasively the function of bioreactors, many of these devices have yet to be fully validated. The stainless steel systems have long developed protocols for validation and, thus, are favored by Big Pharma for its assuredness. While there are still some glitches, most of the functions of disposable bioreactors can be readily monitored and validation protocols run to prove their reproducibility. This is one area where innovation has begun to payback heavily. There are now several new concepts working here: the use of fluorescence to measure parameters from a distance, and the same is done with near-IR spectra. Every parameter that is of value in PAT will be monitored in a totally noninvasive manner in the future, and this will allow manufacturers to place their products in the market sooner.

Leachables

The unpredictability of the leaching of chemicals from plastic materials will continue to hound the industry in the near future. The issue is less important in upstream processing and for all steps before purification, yet the barrage of publications, not all of which are unbiased, regarding these problems has not helped the case of equipment manufacturers. The suppliers have responded by providing excellent support to clients by providing data on the leachables in their products. Equipment manufacturers have also teamed up with CROs who are willing to conduct full GLP studies for just a few thousands of dollars to provide fully qualified data. It is anticipated that there would be no way out of conducting these studies even if parts of the plastic components are replaced with materials such as Teflon or Gore-Tex that do not leach: there will always be a component in the line of components that would make the study essential.

Animal Origins

Responding to the risk of bovine spongiform encephalopathy (BSE) and transmissible spongiform encephalopathy (TSE), companies have begun avoiding animal-derived stearates used in polymerization reactions. To comply with worldwide requirements, the use of animal-derived components in the manufacture of plastics will cease.

The Stainless Challenge

Single-use technology has been used in the biotechnology industry for almost a decade, yet very few manufacturers have switched or retooled their facilities to solely single-use equipment. Companies with existing conventional infrastructure in stainless steel, and considering the associated asset depreciation, are resistant to changing to single-use equipment. There is also a lack of familiarity with the renovation and the cost of building a new facility. With stainless steel, the resistance occurs with facilities that already have stainless steel equipment that is validated: they do not want to implement disposable-type operations.

If there are validated preexisting stainless steel systems within a facility where some process operations could go completely disposable (e.g., chromatography steps, tangential flow filtration, and sterile filtration), there is more resistance to change to disposables systems.

Standardization

As disposable components found their way from downstream to upstream, from filters and buffer preparation to bioreactors, equipment suppliers developed their own standards and specification of things such as tube connectors, rating of sterilizing filters, and, above all, the composition of films. Thus, today, if a single component in a chain is substituted, this may require fresh leachable studies. System designs, such as the type of manifolds, can be varied and, unless these are standardized, it will be difficult to mix and match suppliers of disposable components. Quantifying the economic benefits of single-use systems has also been challenging despite the obvious advantages. The Bioprocess System Alliance (BPSA), Washington D.C., recently published the "Roadmap to Implementation of Single-Use Systems" (published in the April 2010 supplement of *BioProcess International* (BPI), and is available on the BPSA and BPI websites). When one tries to quantify the monetary cost and savings for a single-use process or facility, it can be a difficult exercise because many factors involved in the cost of a stainless steel process or factory have yet to be captured. One example is the cost of making WFI. The savings in WFI is quite significant with disposables; but although it is relatively easy to quantify water use, many companies find it difficult to determine the actual cost to make WFI. Economics can be driven more by capital and overhead use than operating cost. It is quite clear, however, that there are savings to be had, and decisions for single-use systems are being made even when cost savings analyses are unresolved. It is anticipated that in the future the bioprocessing industry will develop some unified standards to allow the integration of multivendor supplies.

Upstream

In the future, biological drugs will be manufactured very differently. From the possibility of cell-free expression systems, such as Invitrogen's Expressway (www.invitrogen.com), to solid-state chemical synthesis, it is possible that one would no longer need to handle live cells and organisms and, thus, removing many challenges such as viral clearance and structure variability in the future. However, these systems are unlikely to be in mainstream production for several decades, if not longer.

Compliance

The basic requirements of keeping cross-contamination out of the system will become so onerous that manufacturers will be left with no choice but to

adopt disposable systems. As long as the cost of changeover, cleaning validation, and compliance stays within reach, few manufacturers have an incentive to mothball their stainless steel systems. However, as biogeneric drugs
begin entering the market, innovators will feel a real pinch and may adopt
the new way of manufacturing these drugs.

High-Expression Cell Lines

Many advances have been made in the titer obtained from animal cell cultures. The magical level of 10 g/L is being reached, which is what is needed
to cut down the entire production operation in the upstream area. As a
result, future upstream facilities will shrink substantially. Many facilities
with designs based on 1 to 2 g/L titers will become obsolete as they would
bring a disconnect between upstream and downstream processing. A 2,000
L disposable bioreactor would be able to do what a 20,000 L bioreactor of the
past did. The size of bioreactors will inevitably shrink.

Flexible Factories

To anticipate future needs, manufacturing facilities need to move away
from large-scale, stick-built-fixed configuration designs to those that will
be less expensive and easier to build. Clean spaces will be optimized and
can easily be expanded and contracted. Interiors will be highly configurable with utility panels and portable equipment to accommodate product
mixes and different production volumes. Innovations are also needed to
comply with the newer or upcoming requirements of the FDA and EMEA
in ensuring viral clearance, and this will require adopting unorthodox
heating, ventilation, and air conditioning (HVAC) systems instead of
physically separating the activities. A bioprocess facility with this mobile
warehouse design and disposable applications will offer several competitive advantages for manufacturing. Quick construction and setup is particularly advantageous for vaccine production in response to pandemic
or bioterrorist threats at desirable geographical locations. Because capital
investment is relatively lower than for conventional facilities (and operations are also simpler with the application of disposable technologies),
lower fixed costs and less labor will be required to maintain and operate
these new facilities.

Small Companies

It is well established that no matter how Big Pharma reduces its cost, it can never compete with smaller companies on the cost of production: with disposable systems, many small companies will be able to offer the quality and quantity of products that are needed by Big Pharma, which means it would be prudent for Big Pharma to outsource manufacturing of their biological APIs. The shift that took place 40 years ago when the manufacturing of chemical APIs was outsourced is about to take place for biological drugs, and smaller, efficient companies built using this max-dispo philosophy will be in high demand.

Unitary Systems

The predictions about the future of biological manufacturing are evolutionary, as demand and supply meet and as the awareness of the cost reductions and need for cost reductions collide. However, the real changes in the universe of biological manufacturing will come from deconstructing the locked-in paradigms of manufacturing that continue to survive. The talk about upstream and downstream stages continues as if they are inextricably inseparable. A disruptive technology breakthrough will take advantage of the flexibility of disposable systems and combine upstream and downstream stages. A typical process of monoclonal antibody production should be examined. A cell line is introduced in a flexible bag to grow the culture and, when it reaches a certain confluence, allows it to express the target protein, all of which can be done swiftly in a disposable container such as the Wave Bioreactor or the newly introduced MayaBioReactors (www.mayabio.com). The next step is to remove the cells by filtering them through a 0.22 μm filter, reduce the volume of media to about one-tenth of the volume, and then load the remaining solution into a column, such as an ion-exchange column for erythropoietin, for further purification. As long as the media volumes are tens to hundreds of liters, these steps are manageable, but when one enters thousand liter systems, the time and cost for these process steps become prohibitive. Furthermore, the delays in processing and the force applied on the target proteins in the solution reduce the yields. The question arises that if the purpose is to isolate the target protein from a large mass of culture media, why are components such as cells and fluids being removed from the media? Why not just pick up the target protein using a resin or a combination of resins, and drain out the balance of media? The resin can be a specifically made resin or a mixture of resins for nonspecific binding. Systems to remove resins may include introducing the resin in filter pouches that can be physically removed from the bioreactors after equilibration and directly transferred into columns. Suddenly, there is a unit process that replaces three processes. One can even go further and use the flexible bag as the first purification column. These innovations are in the offing and are a subject of several U.S. patents filed by MayaBio Company (www.mayabio.com).

The use of bioreactors for bacterial cultures poses another set of challenges that will eventually be resolved, wherein chemical methods will be used to disrupt bacteria in the culture medium, additional chemical means to unfold the inclusion bodies and refold proteins, all in the same bag where the fermentation process takes place. In the end, a resin capable of capturing partially folded or unfolded protein will be used to separate the target protein, discarding the balance of the mixture in the disposable bag.

Biosafety

The industry has historically taken a very cautious approach toward the disposal of GMOs and many of them overbuilt their facilities to decontaminate objects. In those cases where the GMOs fall in the exempt category of biosafety, such as recombinant CHO cells, these may be discharged in the municipal sewer systems. In the past, the industry became overconcerned about releasing recombinant agents into the atmosphere, but if an entity is not infectious or is unable to survive outside of a specific environment, these fall under an exempt qualification as described in Appendices C and K of the NIH biosafety guidelines. However, the manufacturers must also consult with local requirements of biosafety compliance and find out if GLSP guidelines apply to them or not.

Autoclaves

Autoclaves will not be needed: the GMO contact surfaces and effluents can all be decontaminated by chemical means. Autoclaves are also used to sterilize multiuse products, media, buffers, connectors, and other items that come in contact with the product; the use of disposable systems makes them redundant. The autoclave is the single most energy-consuming and energy-dissipating item using high-quality water in very large quantities. The heat dissipated by an autoclave upsets the balance of HVAC calculations and the facilities need to be sized to accommodate frequent use of autoclaves. Additionally, autoclaves add perhaps millions of dollars of capital investment. In those instances where decontamination is required, this would mainly be a chemical process and made part of the discharge from the bags, removing the need for a holding tank and allowing direct discharge in sewer systems.

SIP/CIP

There will be nothing that will require cleaning or sterilization prior to use—all systems would come presterilized and prefilled where necessary. This is one single source of the largest use of water, and the savings would accrue not only in the quantity of water but all systems that lead to the availability of pure steam. The cost of SIP/CIP is always underestimated in cost calculations; besides the use of water, it should include the systems that need maintenance to provide clean steam.

Distilled Water Loops

It is customary for the industry to have a stainless steel loop of distilled water run at 80°C. The preparation of distilled water involves water purification, deionization, boiling water, condensing water, storing distilled water, maintaining distilled water loops by passivation, sanitizing the loops, and testing the ports for compliance with the quality of water discharged. The industry requires only purified water, and this can be readily provided by the double RO/EDI systems (see the manufacturing facility of Therapeutic Proteins, Inc. (www.theraproteins.com) for an example) once the needs are reduced by almost 90% with the elimination of the autoclave and SIP/CIP systems.

Low Ceiling Heights

Vertical ceiling heights will not be an impediment, as horizontal bioreactors would replace the current multistory bioreactors, and so the cost of building new facilities will decrease significantly.

Modular Systems

Modular systems of connecting disposable bags will obviate the need for scale-up and allow custom manufacturing of variable batch sizes. Only mixing validation will be required, which can be readily demonstrated.

Gentle Mixing

Upstream systems have historically been very rigorous; the new findings are that all cells and organisms grow much better when they are allowed to grow in colonies, which means providing gentler mixing systems. This profile is well delivered in 2D bag bioreactors only as 3D still require propellers and mixing systems similar to what is used in the stainless steel bioreactors.

2D Bags

Disposable bags will mainly be of the 2D type as this is the cheapest form, and the hybrid disposables that are designed to provide a liner to the existing stainless steel-type bioreactors with disposable mixing attachments will fade because of their high cost without any additional benefits. Every type of cell and organism will be adequately grown in 2D plastic bags with innovations in the designs of the bags to meet the growth requirements (see www. mayabio.com). The problem with 2D bags lies in the difficulties encountered when handling larger volumes and also to hold. For example, the only large commercial 2D bag bioreactor comes from GE (the Wave Bioreactor); beyond the 500 L media, the bags would not be able to hold the weight as it is rocked. However, recent filings with the U.S. Patent and Trademark Office show that a new class of stationary bioreactors will soon be appearing where the 2D bag is kept stationary (thus, eliminating any size constraints) and the aeration is provided through proprietary ceramic spargers. The 3D bag technologies as currently promoted by all major equipment manufacturers require disposal mixing systems preinstalled and that adds substantial ongoing costs, while the starting equipment also adds substantial capital investment. Today, the cost of a disposable 3D bioreactor is just about the same as the stainless steel bioreactors. The use of 2D stationary bags will cut down the capital cost by more than 50%.

Fluoropolymer Bags

Fluorinated polymers such as Teflon (polytetrafluoroethylene) do not contain any extractables or leachables and will ideally be the choice for making disposable bags. While the science to do this is available today, the disposable bags industry remains locked in with traditional polyethylene bags with its chemical additives, mainly because they have already made huge investments in the equipment to make these bags and partly because of the ease how standard plastic bags are made. For example, a temperature of 400°F is sufficient to seal plastic bags, and a temperature of 800°F is need to handle Teflon: the current manufacturing technology would require a total revamp to accommodate Teflon bags. Other problems in the use of fluoropolymer compounds are their permeability to gases, and one form of Teflon, Gore-Tex, even allows moisture to escape but retains water inside. To overcome these problems, MayaBio (www.mayabio.com) has established a proof of principle of using a "double bagging" system whereby a Teflon bag is inserted inside a plastic bag and the culture media comes in contact with Teflon only. Studies are needed to confirm that the surface characteristics of Teflon will not adversely affect the growth of culture media. However, the basic principle has been earlier proven using animal cells and using Teflon bags to produce working culture. If Teflon or Telfon-coating plastic becomes readily available, this would eliminate one of the most significant hurdles in the acceptance of disposable bioreactors as the regulatory requirements of proving the safety of contact materials would be removed.

Protein Capture

Target protein-specific resins will be developed that will allow fewer steps in downstream purification. The example of Protein A as a capture resin for monoclonal antibodies will be extended to other types of resins, and affinity chromatography would become the most widely used method for downstream processing. U.S. Patent (xx) reports a new kind of bioreactor that captures the target molecules as they are produced using an electro-deionization step. Other recent filings with the U.S. Patent and Trademark Office show that it will be possible in the future to remove target proteins directly from the bioreactor and even perform purification using the bioreactor as the downstream column. Details are available at www.mayabio.com.

Downstream Processing

Molecule-specific facilities and equipment will prove to be more cost-effective than using a disposable system. The current attempts to make the tubing in chromatography systems disposable will not be needed if the equipment is dedicated to each molecule. With the development of better resins and the partial purification achieved in the bioreactor, it will be more cost-effective to dedicate equipment to molecules.

Closed Systems

Completely closed systems wherein buffers, media, protein capture, and initial purification remain enclosed will allow the use of environment conditions that are less expensive to maintain, for example, the use of a 500K facility instead of a 100K facility currently recommended for upstream processing.

Molecule-Specific Facilities

Molecule-specific facilities will become the norm to reduce the largest burden of cross-contamination validation. This will become possible as the capital cost of establishing a facility reduces substantially. The focus will be on smaller facilities that operate in a self-contained environment.

Max-Dispo Concept

While the use of disposable components has obvious advantages, with molecule-specific facilities and substantially reduced sizes of production systems, disposable items will be used where needed and not necessarily just because disposable items are available. For example, buffer preparation is a noncontaminating exercise, and there is no need to use disposable stirrers: just the containers need to be disposable. The industry will evolve into integrating a realistic approach regarding what should be disposable.

Leachables/Extractables

This controversy will be ended by the development of plastic films that are inherently free from additives. In addition, through regulatory approval of applications where products are mainly made using disposable systems, the FDA/EMEA will establish the safety of plastic. Currently, no clear guidelines exist from regulatory agencies, leaving a large burden on manufacturers. This will be reduced by the agencies issuing specific guidelines for the use of disposable systems in bioprocessing.

Multipurpose Disposable Bioreactors

The greatest hurdle in providing sufficient oxygen for the ever-higher expressing cell lines has been removed by novel sparging systems to deliver oxygen mass transfer capacity of 10 mmol/L/h ($K_La = 50/h$) when 50×10^6 cells/mL are cultivated. For microbial systems, like an *E. coli* fermentation at 50 g/L dry cell weight, the required mass transfer capacity has to be 200 mmol/L/h ($K_La > 800/h$) or even higher. More advances will be made in the use of disposable bags that will be available with integrated sensors and equipped with connections for feed, inoculums, sampling, and with gas inlet and exhaust gas filters.

In conclusion, the disposable technology of today will prove to be a "game changer" in the field of bioprocessing, making it easier to develop new drugs and to conduct research in the emerging field of stem cells and gene therapy. The driving force behind this change will be the cost savings realized by the industry in making this switch even though it may mean discarding a heavy infrastructure of manufacturing biological drugs.

Bibliography

AAMI TIR17. Radiation Sterilization—Material Qualification; www.aami.org; http://webstore.ansi.org/ansidocstore/product.asp?sku=AAMI+TIR17%3A1997.

AAMI TIR33: Sterilization of Health Care Products—Radiation, Substantiation of a Selected Sterilization Dose—Method VDmax; http://www.aami.org; 2005 http://www.webstore.ansi.org/ansidocstore/product.asp?sku=AAMI+TIR 33%3A2005.

Abdullah, G. 2003. Making the Most of a Powerful Nuisance. *Penn State Agriculture Magazine*: 17–19; http://www.aginfo.psu.edu/psa/f2003/nuisance.html.

AC Engineering. 2010. Disposable Pumps—Product description. Available: http://www.acengineering.co.il/index.aspx?id=2152.

Adam, E., Sarrazin, S., Landolfi, C., Motte, V., Lortat-Jacob, H., Vassalle, P., and Delehedde, M. 2008. Efficient long-term and high-yielded production of a recombinant proteoglycan in eukaryotic HEK293 cells using a membrane-based bioreactor. *Biochem. Biophys. Res. Commun.* 369: 297–302.

Advanced Scientifics. 2010. Three60 Sampling System—Product Description. Available: http://www.asi360.com.

Agalloco, J., Akers, J., and Madsen, R. 2007. Choosing technologies for aseptic filling: "Back to the future, forward to the past?" *Pharm. Eng.* 27: 8–16.

Agalloco, J. and Akers, J. 2008. Sterile product manufacture. In Gad, S.C. (ed.), *Pharmaceutical Manufacturing Handbook-Production and Processes*. Hoboken, NJ: John Wiley & Sons, 99–136.

Albis Technologies AG. 2010. Zurich, Switzerland.

Albrecht, W., Fuchs, H., and Kittelmann, W. (eds.). 2002. *Nonwoven Fabrics*. Weinheim: Wiley-VCH.

Aldington, S. and Bonnerjea, J. 2007. Scale-up of monoclonal antibody purification processes. *J. Chromatogr. B* 848: 64–78.

Altaras, G.M., Eklund, C., Ranucci, C., and Maheswari, G. 2007. Quantitation of lipids with polymer surfaces in cell culture. *Biotechnol. Bioeng.* 96: 999–1007.

American Chemistry Council. PlasticsResource.com: Information on Plastics and the Environment; http://www.plasticsresource.com/s_plasticsresource/index.asp.

Anderlei, T., Cesana, C., De Jesus, M., Kühner, M., and Wurm, F. 2009. Shaken Bioreactors Provide Culture Alternative. GEN 29. Available at: http://www.genengnews.com/issues/articleindex.aspx.

Anders, K.D., Akhnoukh, R., Scheper, T., and Kretzmer, G. 1992. Culture fluorescence measurements for the monitoring and characterization of insect cell cultivation in bioreactors. *Chemie Ingenieur Technik* 64 (6): 572–573.

Anders, K.D., Wehnert, G., Thordsen, O., Scheper, T., Rehr, B., and Sahm, H. 1993. Biotechnological applications of fiber-optic sensing—Multiple uses of a fiber-optic fluorometer. *Sens. Actuators B Chem.* 11 (1–3): 395–403.

Anderson, K.P., Lie, Y.S., Low, M.A., Williams, S.R., Fennie, E.H., Ngyen, T.P., and Wurm, F.M. 1990. Presence and transcription of intracisternal A-particle related sequences in CHO cells. *J. Virol.* 64: 2021–2032.

Anicetti, V. 2009. Biopharmaceutical processes: A glance into the 21st century. *BioProcess Int.* 7(4): S4–S11.

ANSI/AAMI ST67: Sterilization of Health Care Products—Requirements for Products Labeled "Sterile"; http://www.ansi.org.

ANSI/AAMI/ISO 11137-1: Sterilization of Health Care Products—Radiation, Part 1: Requirements for Development, Validation and Routine Control of a Sterilization Process for Medical Devices; http://www.iso.org. 2006.

ANSI/AAMI/ISO 11137-2: Sterilization of Health Care Products—Radiation, Part 2: Establishing the Sterilization Dose; http://www.iso.org. 2006.

ANSI/AAMI/ISO 11137-3: Sterilization of Health Care Products—Radiation, Part 3: Guidance on Dosimetric Aspects; http://www.iso.org. 2006.

ANSI/AAMI/ISO 11737-1: Sterilization of Medical Devices—Microbiological Methods, Part 1: Determination of a Population of Microorganisms on Products; http://www.iso.org. 2006.

ANSI/AAMI/ISO 11737-2: Sterilization of Medical Devices—Microbiological Methods, Part 2: Tests of Sterility Performed in the Validation of a Sterilization Process (first edition); http://www.iso.org. 1998.

Arnold, S.A., Gaensakoo, R., Harvey, L.M., and McNeil, B. 2002. Use of at-line and in-situ near-infrared spectroscopy to monitor biomass in an industrial fed-batch Escherichia coli process. *Biotechnol. Bioeng.* 80 (4): 405–413.

Article 16 of and Annex II to Council Directive 1999/31/EC. Criteria and Procedures for the Acceptance of Waste at Landfills. European Union, 1999; http://www.eur-lex.europa.eu/smartapi/cgi/sga_doc?smartapi!celexplus!prod!CELEXnumdoc&lg=EN&numdoc=32003D0033.

Arunakumari, A., Wang, J.M., and Ferreira, G. 2007. Alternatives to protein A: Improved downstream process design for humanly monoclonal antibody production. *BioPharm Int.* 2: 36–40.

Arunakumari, A. 2006. Integrating high titer cell culture processes with highly efficient purification processes for the manufacturing of human monoclonal antibodies. IBC Conference, San Francisco, CA, November.

ATMI Life Sciences. 2008. Mixing & resuspension of high powder loads using a magnetic mixer. Application note. Available from ATMI Life Sciences.

ATMI Life Sciences. 2008. Mixing a diatomaceous earth slurry using the Jet-Drive mixer. Application note. Available from ATMI Life Sciences.

ATMI Life Sciences. 2008. Mixing a diatomaceous earth slurry using a Pad-Drive mixer. Application note. Available from ATMI Life Sciences.

ATMI Life Sciences. 2008. Mixing a diatomaceous earth slurry using the WandMixer. Application note. Available from ATMI Life Sciences.

ATMI Life Sciences. 2008. Mixing of high powder loads using a LevMixer. Application note. Available from ATMI Life Sciences.

ATMI Life Sciences. 2008. Particle generation in the Jet-Drive mixer. Application note. Available from ATMI Life Sciences.

ATMI Life Sciences. 2008. Particle generation in the Magnetic Mixer. Application note. Available from ATMI Life Sciences.

ATMI Products. 2009. Catalog.

ATV. 1996. Abwasser aus gentechnischen Produktionsanlagen und vergleichbaren Einrichtungen. Hennef: DWA.

ATV. 2001. Abwasser aus Krankenhaäusern und anderen medizinischen Einrichtungen. Hennef: DWA.

Aunins, J.B., Bibila, T.A., Gatchalian, S., Hunt, G.R., Junker, B.H., Lewis, J.A., Seifert, D.B., Licari, P., Ramasubramanyan, K., Ranucci, C.S., Seamans, T.C., Zhou, W., Waterbury, W., Buckland, B.C. 1997. Reactor development for the hepatitis A vaccine VAQTA. In: Carrondo, M.J.T., Griffiths, B., and Moreira, J.L.P. (eds.), *Animal Cell Technology: From Vaccine to Genetic Medicine*. Kluwer, Dordrecht, pp 175–183.

Bail, P., Crawford, B., and Lindström, K. 2009. 21st century vaccine manufacturing. *BioProcess Int.* 4: 18–28.

Bandrup, J. 1999. Verfahrenswege der Kunststoffverwertung aus ökonomischer und ökologischer Sicht. In K. Wiemer and N. Kern (eds.), *Bio- und Restabfallbehandlung*. Witzenhausen: Witzenhausen Institute.

Barbaroux, M. and Sette, A. 2006. Properties of materials used in single-use flexible containers: Requirements and analysis. Available: http://biopharminternational. findpharma.com/biopharm/article/articleDetail.jsp?id=423541&sk=&date=&p ageID=7.

Barnoon, B. and Bader, B. 2008. Lifecycle cost analysis for single-use systems. *BioPharm Int.*

Bartolome, A.J, Ulber, R., Scheper, T., Sagi, E., and Belkin, S. 2003. Genotoxicity monitoring using a 2D-spectroscopic GFP whole cell biosensing system. *Sens. Actuators B Chem.* 89 (1–2): 27–32.

Bean, B., Matthews, T., Daniel, N., Ward, S., Wolk, B. 2008. Guided wave radar at Genentech: A novel technique for noninvasive volume measurement in disposable bioprocess bags: GWR may be a cheaper, more practical alternative to traditional methods. Available: http://www.pharmamanufacturing.com/ articles/2008/185.html.

Becker, T., Mitzscherling, M., and Delgado, A. 2002. Hybrid data model for the improvement of an ultrasonic-based gravity measurement system. *Food Control* 13 (4–5): 223–233.

Beeksma, L.A. and Kompier, R. 1995. Cell growth and virus propagation in the costar cell cube system. In: Beuvery, E.C., Griffiths, J.B., and Zeijlemaker, W.P. (eds.), *Animal Cell Technology: Developments towards the 21st Century*. Kluwer, Dordrecht, 661–663.

Behme, S. 2009. *Manufacturing of Pharmaceutical Proteins*. Weinheim: Wiley-VCH Verlag GmbH & Co. KGaA.

Behme, S. 2009. Production facilities. In: Behme, S. (ed.) *Manufacturing of Pharmaceutical Proteins*. Wiley VCH, Weinheim, 227–275.

Bender, J. and Wolk, B. 1998. Putting a spin on CHO harvest. Centrifuge technology development. ACS Meeting, Boston, MA.

Bennan, J. et al. 2002. Evaluation of extractables from product-contact surfaces. *BioPharm. Int.* 15(12): S22–S34.

Bentebibel, S., Moyano, E., Palazón, J., Cusidó, R.M., Bonfill, M., Eibl, R., Piñol, M.T. 2005. Effects of immobilization by entrapment in alginate and scale-up on paclitaxel and baccatin III production in cell suspension cultures of Taxus baccata. *Biotechnol. Bioeng.* 89: 647–655.

Bergveld, P. 1970. Development of an ion-sensitive solid-state device for neurophysiological measurements. *IEEE Trans. Biomed. Eng.* 19: 70–71.

Bergveld, P. 2003. Thirty years of ISFETOLOGY: What happened in the past 30 years and what may happen in the next 30 years. *Sens. Actuators B Chem.* 88 (1): 1–20.

Bernard, F. et al. 2009. Disposable pH sensors. *BioProcess Int.* 7(1): S32–S36.

Bestwick, D. and Colton, R. 2009. Extractables and leachables from single-use disposables. *BioProcess Int*. 7(1): S88–S94.

Beyond Borders Global Biotechnology Report 2007. Zurich/Boston: Ernst & Young.

Bioengineering. 2001. Aerosolfree Hi Containment sampling system. *Bioengineering-Culture in Hygienic Design*, 37–40.

Bioengineering. 2001. Aseptic connection of two sterile spaces. *Bioengineering-Culture in Hygienic Design*, 21–36.

BioProcess Systems Alliance (BPSA) Disposal subcommittee. 2008. Guide to disposal of single-use bioprocess systems. *BioProcess Int*. 6(5): S24–S27.

BioWorld Snapshots. 2009. Biotechnology Products on the Markets Since 1982, *BioWorld Today*, Atlanta.

Boehl, D., Solle, D., Hitzmann, B., and Scheper, T. 2003. Chemometric modelling with two-dimensional fluorescence data for Claviceps purpurea bioprocess characterization. *J. Biotechnol*. 105 (1–2): 179–188.

Boehm, J. and Bushnell, B. 2007. Providing sterility assurance between stainless steel and single-use systems. *BioProcess Int*. 5(4): S66–S71.

Booth, A. 2007. *Radiation Sterilisation, Validation and Routine Operations Handbook*. River Grove, IL: Davis Healthcare International Publishing.

Bosch Packaging Technology 2008. Available: http://www.pharmaceuticalonline.com/article.mvc/New-Prevas-Disposable-Dosing-System-For-Filli-0001?VNETCOOKIE=NO.

Boss, J. 1986. Evaluation of the homogeneity degree of a mixture. *Bulk Solids Handling* 6 (6): 1207–1215.

Brecht, R. 2009. Disposable bioreactors—Maturation into pharmaceutical glycoprotein manufacturing. In Eibl, D. and Eibl, R. (eds.), *Disposable Bioreactors, Series: Advances in Biochemical Engineering/Biotechnology*, Vol. 115. Berlin; Heidelberg: Springer, 1–31.

British Standards Online 2007. EN 13431 Packaging—Requirements for Packaging Recoverable in the Form of Energy Recovery, Including Specification of Minimum Inferior Calorific Value; http://www.bsonline.si-global.com.

Brorson, K., Krejci, S., Lee, K., Hamilton, E., Stone, K., and Xu, Y. 2003. Bracketed generic inactivation of rodent retroviruses by low pH treatment for monoclonal antibodies and recombinant proteins. *Biotechnol. Bioeng*. 82: 321–329.

Brorson, K. 2006. CDER/FDA. Virus filter validation and performance. Recovery of Biological Products XII, Phoenix, AZ.

Brough, H., Antoniou, C., Carter, J., Jakubik, J., Xu, Y., and Lutz, H. 2002. Performance of a novel Viresolve NFR virus filter. *Biotechnol. Prog*. 18: 782–795.

Brown, L.F. and Mason, J.L. 1996. Disposable PVDF ultrasonic transducers for non-destructive testing applications. *IEEE Trans. Ultrason. Ferroelectr. Freq. Control* 43 (4): 560–568.

Bruce, M.P., Boyd, V., Duch, C., and White, J.R. 2002. Dialysis-based bioreactor systems for the production of monoclonal antibodies—Alternatives to ascites production in mice. *J. Immunol. Methods* 264: 59–68.

Büchs, J., Maier, U., Milbradt, C., and Zoels, B. 2000. Power consumption in shaking flasks on rotary shaking machines: I. Power consumption measurements in unbaffled flask at low viscosity. *Biotechnol Bioeng* 68: 589–593.

Burrill, G.S. 2010. *Biotech IPOs Completed 1996–2009*. San Francisco: Burrill & Company.

Bush, L. 2008. Disposal of disposables. *BioPharm Int*. 7: 12.

Cabatingan M. 2005. Impact of virus stock quality on virus filter validation: A case study. *BioProcess Int* I 1: 39–43.

Cai, K., Gierman, T.M., Hotta, J., Stenland, C.J., Lee, D.C., Pifat, D.Y., and Petteway, S.R. 2005. Ensuring the biologic safety of plasma-derived therapeutic proteins. *BioDrugs* 19: 79–96.

Cappia, J.-M. and Holman, N.B.T. 2004. Integrating single-use disposable processes into critical aseptic processing operations. *BioProcess Int.* 2(4); http://www.bioprocessintl.com/multimedia/archive/00078/0209su12_78577a.pdf.

Cardona, M. and Allen, B. 2006. Incorporating single-use systems in biopharmaceutical manufacturing. *BioProcess Int.* 4(4, supplement): 10–14.

Castellarnau, M., Zine, N., Bausells, J., Madrid, C., Juárez, A., Samitier, J., and Errachid, A. 2007. Integrated cell positioning and cell-based ISFET biosensors. *Sens. Actuators B Chem.* 120 (2): 615–620.

Castillo, J. and Vanhamel, S. 2007. Cultivating anchorage-dependent cells. *Genet. Eng. Biotechnol. News* 16: 40–41.

Castillo, J. and Vanhamel, S. 2007. Cultivating anchorage-dependent cells. *GEN* 27: 40–41.

Cellexus Biosystems. 2006. Aseptic Sampling from Bioreactors. Available: http://www.proteigene.com/pdf/Cellexussampler.pdf.

Center for Drug Evaluation and Research. Guidance for Industry: Container Closure Systems for Packaging Human Drugs and Biologics—Chemistry, Manufacturing, and Controls Documentation. US Food and Drug Administration: Rockville, MD; 1999 http://www.fda.gov/cder/guidance/1714fnl.htm.

Chapter <87> Biological Reactivity Tests, In Vitro, USP 30. United States Pharmacopeial Convention: Rockville, MD. 2007.

Chapter <88> Biological Reactivity Tests, In Vivo, USP 30. United States Pharmacopeial Convention: Rockville, MD. 2007.

Charles I., Lee J., Dasarathy Y. 2007. Single-use technologies—A contract biomanufacturer's perspective. *BioPharm Int.* 20 (Nov. Suppl.): 31–36.

Cheng, L.J., Li, J.M., Chen, J., Ge, Y.H., Yu, Z.R., Han, D.S., Zhou, Z.M., and Sha, J.H. 2003. NYD-SP16, a novel gene associated with spermatogenesis of human testis. *Biol. Reprod.* 68 (1): 190–198.

Chmiel, H. 2006. Bioreaktoren. In: Chmiel H (ed.), *Bioprozesstechnik*. Elsevier, München, 195–215.

CHMP/CVMP. Guideline on Plastic Immediate Packaging Materials. European Medicines Evaluation Agency: London, UK; 2005 http://www.emea.europa.eu/pdfs/human/qwp/435903en.pdf.

Chovelon, J.M., Fombon, J.J., Clechet, P., Jaffrezic-Renault, N., Martelet, C., Nyamsi, A., and Cros, Y. 1992. Sensitization of dielectric surfaces by chemical grafting: Application to pH ISFETs and REFETs. *Sens. Actuators B Chem.* 8 (3): 221–225.

Clutterbuck, A., Kenworthy, J., and Lidell, J. 2007. Endotoxin reduction using disposable membrane adsorption technology in cGMP manufacturing. *BioPharm. Int.* 20: 24–31.

Code of Federal Regulations. 2009. Title 21 CFR 211. Current good manufacturing practice for finished pharmaceuticals. National archives and records administration. Office of the Federal Register of the USA.

Code of Federal Regulations. 2009. Title 21, part 210. Current good manufacturing practice for finished pharmaceuticals. National archives and records administration. Office of the Federal Register of the USA.

Colder Products. 2010. Steam-Thru Connections—Product description. Available: http://www.colder.com/Products/SteamThruConnections/tabid/740/Default.aspx.

Cole, G. 1998. Pharmaceutical Production Facilities: Design and Applications. Pharmaceutical Science Series. London: Taylor & Francis, Informa Health Care.

Cole Parmer. 2010. Silicon Tubing—Product description. Available: http://www.coleparmer.com/Catalog/product_view.asp?sku=9610500.

Colton, R. 2007. Recommendations for extractables and leachables testing—Part 1. *BioProcess Int*. 11: 36–44.

Colton, R. 2007. The extractables and leachables subcommittee of the bio-process systems alliance. Recommendations for extractables and leachables testing, part 1: Introduction, regulatory issues and risk assessment. *BioProcess Int*. 12: 36–44.

Colton, R. 2008. Recommendations for extractables and leachables testing—Part 2. *BioProcess Int*. 1: 44–52.

Colton, R. 2008. The extractables and leachables subcommittee of the bio-process systems alliance. Recommendations for extractables and leachables testing, part 2: Executing a program. *BioProcess Int*. 1: 44–52.

Committee for Proprietary Medicinal Products (CPMP). 1996. Mark for guidance on virus validation studies: The design, contribution and interpretation of studies validating the inactivation and removal of viruses. CPMP/BWP/268/95.

Committee for Proprietary Medicinal Products (CPMP). 2000. Annex to Note for Guidance on Development Pharmaceutics: Decision Trees for the Selection of Sterilisation Methods. London: European Medicines Evaluation Agency.

Computational-fluid-dynamics (CFD) analysis of mixing and gas-liquid mass transfer in shake flasks. *Biotechnol. Appl. Biochem*. 41: 1–8.

CPMP. 1996. Note for guidance on virus validation studies. The design, continuation and interpretation of studies validating the inactivation and removal of viruses. CPMP/BWP/268/95.

CPMP. 1997. Mark for guidance on quality of biotechnology products, viral safety evaluation of biotechnology products derived from cell lines of humanly or animal origin. CPMP/I/295/9S. London.

CPT Consolidated Polymer Technologie, Inc. 2002. Material Comparison. Brochure. Available: http://www.stiflow.com/CFIex-Tubing-Material-Comparison.pdf.

Croughan, G. 2010. Beyond Just High Titers: The Future of Cell Line Engineering. IBC Cell Line Development and Engineering Conference, San Francisco, CA.

Curling, J. and Gottschal, U. 2007. Process chromatography: Five decades of innovation. *BioPharm. Int*. 21: 70–94.

Curtis, S., Lee, K., Blank, G.S., Brorsen, K., and Xu, Y. 2003. Generic/matrix evaluation of SV40 clearance by anion exchange chromatography in flow-through mode. *Biotechnol. Bioeng*. 84: 179–186.

Curtis, W.R. 1999. Achieving economic feasibility for moderate-value food and flavour additives. In Fu, T., Singh, G., and Curtis, W.R. (eds.), *Plant Cell and Tissue Culture for the Production of Food Ingredients*. New York: Kluwer Academic, 225–236.

Curtis, W.R. 2004. Growing cells in a reservoir formed of a flexible sterile plastic liner. United States Patent 6709862B2.

D'Aquino, R. 2006. Bioprocessing systems go disposable. *Chem. Eng. Prog*. 102 (5): 8–11.

Dancette, O.P., Taboureau, J.L., Tournier, E., Charcosset, C., and Blond, P. 1999. Purification of immunoglobulins G by protein A/G affinity membrane chromatography. *J. Chromatogr. B* 723: 61–68.

Danckwerts, P.V. 1953. Continuous flow systems: Distribution of residence times. *Chem. Eng. Sci.* 2: 1–13.

Davis, J.M. 2007. Hollow fibre cell culture. In Pörtner, R. (ed.), *Animal Cell Biotechnology: Methods and Protocols, Series Methods in Biotechnology*, Vol. 24. Totowa, NJ: Humana Press, 337–352.

Davis, J.M. 2007. Hollow fibre cell culture. In: Pörtner, R (ed.), *Animal Cell Biotechnology: Methods and Protocols*. Humana, Totowa, 337–352.

Davis, J.M. 2007. Systems for cell culture scale-up. In Stacey, G. and Davis, J.M. (eds.), *Medicines from Animal Cell Culture*. Chichester, UK: John Wiley & Sons, 145–171.

Davis, R.M. and Taylor, G. (1950). The mechanics of large bubbles rising through liquids in tubes. *Proc. R Soc. Lond. A* 200: 375–392.

DeGrazio, F.L. 2006. The importance of leachables and extractables testing for a successful product launch. Available: http://pharmtech.fmdpharma.com/pharmtech/Article/The-Importance-of-Leachables-and-Extractables-Test/ArticleStandard/Article/detail/482447.

De Jesus, M.J., Girard, P., Bourgeois, M., Baumgartner, G., Kacko, B., Amstutz, H., and Wurm, F.M. 2004. TubeSpin satellites: A fast track approach for process development with animal cells using shaking technology. *Biochem. Eng. J.* 17: 217–223.

De Jesus, M.J. and Wurm, FM. 2009. Medium and process optimization for high yield, high density suspension cultures: From low throughput spinner flasks to high throughput millilitre reactors. *BioProcess Int.* 7 (Suppl. 1): 12–17.

Denton, A., Jones, C., and Tarrach, K. 2009. Integration of large-scale chromatography with nanofiltration for an ovine polyclonal product. *Pharm. Technol.* 1: 62–70.

DePalma, A. 2002. Options for anchorage-dependent cell culture. *Genet. Eng. Biotechnol. News* 22: 58–62.

DePalma, A. 2004. Bioprocessing on the way to total-disposability manufacture. *Genet. Eng. Biotechnol. News* 24 (3): 40–41.

DePalma, A. 2005. Liquid mixing: Solid challenges. Available: http://www.pharmamanufacturing.com/articles/2005/297.html.

DePalma, A. 2006. Bright sky for single-use bioprocess products. *Genet. Eng. Biotechnol. News* 26 (3): 50–57.

Desai, M.A., Rayner, M., Burns, M., and Bermingham, D. 2000. Application of chromatography in the downstream processing of biomolecules. *Methods Biotechnol.* 9: 73–94.

De Wilde, D., Noack, U., Kahlert, W., Barbaroux, M., and Greller, G. 2009. Bridging the gap from reusable to single-use manufacturing with stirred, single-use bioreactors. *BioProcess Int* 7(Suppl 4): 36–42.

DiBlasi, K., Jornitz, M.W., Gottschalk, U., and Priebe, P.M. 2006. Disposable biopharmaceutical processes—Myth or reality? *BioPharm. Int.* 11: 2–10.

Diehl, T. 2006. Application of membrane chromatography in the purification of humanly monoclonal antibodies. Downstream Technology Forum, King of Prussia, PA.

DiMasi, J.A. and Grabowski, HG. 2007.The cost of biopharmaceutical R&D: Is biotech different? *Manage. Decis. Econ.* 28: 469–479.

DIN EN 556. Sterilization of Medical Devices—Requirements for Medical Devices to Be Designated "Sterile"; http://webstore.ansi.org.

DIN ISO 15378. 2007–2010. Primärverpackungen für Arzneimittel—Besondere Anforderungen für die Anwendung von ISO 9001:2000 entsprechend der Guten Herstellungspraxis (GMP) (ISO 15378:2006), October 2007, DIN-Norm.

Ding, W. and Martin, J. 2008. Implementing single-use technology in biopharmaceutical manufacturing: An approach to extractables/leachables studies, Part One—connections and filters. *BioProcess Int.* 6(9): 34–42.

Directive 2000/76/EC of the European Parliament and of the Council on the Incineration of Waste. Off. J. European Communities: 91–111; 2000; http://www.eur-lex.europa.eu/LexUriServ/LexUriServ.do?uri=OJ:L:2000:332:0091:0111:EN:PD.

Directive 2000/76/EC of the European Parliament and of the Council on the Incineration of Waste. Official Journal of the European Communities; 2000http://www.ec.europa.eu/environment/wasteinc/newdir/2000-76_en.pdf.

Dremel, B.A.A. and Schmid, R.D. 1992. Optical sensors for bio-process control. *Chemie Ingenieur Technik* 64 (6): 510–517.

Drugmand, J.C., Havelange, N., Debras, F., Collignon, F., Mathieu, E., and Castillo, J. 2009. Human and Animal Vaccine Production in a New Disposable Fixed-Bed Bioreactor. Available: http://www.artelis.be./uploads/pdf.POSTER%20ARTEFIX%20BD.pdf.

Ducos, J.P., Terrier, B., Courtois, D., and Pétiard, V. 2008. Improvement of plastic-based disposable bioreactors for plant science needs. *Phytochem. Rev.* 7: 607–613.

Ducos, J.P., Terrier, B., and Courtois, D. 2009. Disposable bioreactors for plant micropropagation and cell cultures. In Eibl, D. and Eibl, R. (eds.), *Disposable Bioreactors, Series: Advances in Biochemical Engineering/Biotechnology*, Vol. 115. Berlin; Heidelberg: Springer, 89–115.

Edelmann, W., Arnet, M., Schwarzenbach, H.U., and Stutz, E. 2004. Kunststoffverwertung im Kanton Zug. Zug: ZEBA.

Eibl, R. and Eibl, D. 2002. Bioreactors for plant cell and tisue cultures. In: Oksman-Caldentey, K.M. and Barz, W.H. (eds.), *Plant Biotechnology and Transgenic Plants*. New York: Marcel Dekker, 163–199.

Eibl, R. and Eibl, D. 2006. Design and use of the wave bioreactor for plant cell culture. In Dutta Gupta, S. and Ibaraki, Y. (eds.), *Plant Tissue Culture Engineering, Series: Focus on Biotechnology*, Vol. 6. Dordrecht, The Netherlands: Springer, 203–227.

Eibl, R. and Eibl, D. 2007. Disposable bioreactors for cell culture-based bioprocessing. *ACHEMA Worldwide News* 2: 8–10.

Eibl, R. and Eibl, D. 2007. Disposable bioreactors for inoculum production and protein expression. In Pörtner, R. (ed.), *Animal Cell Biotechnology: Methods and Protocols, Series: Methods in Biotechnology*, Vol. 24. Totowa, NJ: Humana Press, 321–335.

Eibl, R. and Eibl, D. 2008. Application of disposable bag bioreactors in tissue engineering and for the production of therapeutic agents. In Kasper, C., van Griensven, M., and Pörtner, R. (eds.), *Bioreactor Systems for Tissue Engineering*, Vol. 112. Berlin; Heidelberg: Springer, 183–207.

Eibl, R. and Eibl, D. 2008. Bioreactors for mammalian cells: general overview. In Eibl, R., Eibl, D., Pörtner, R., Catapano, G., and Czermak, P. (eds.), *Cell and Tissue Reaction Engineering*. Heidelberg: Springer, 55–82.

Eibl, R. and Eibl, D. 2008. Design of bioreactors suitable for plant cell and tissue cultures. *Phytochem. Rev.* 7: 593–598.

Eibl, R. and Eibl, D. 2009. Disposable bioreactors in cell culture-based upstream processing. *BioProcess Int.* 7 (Suppl. 1): 18–23.

Eibl, R. et al. 2010. Disposable bioreactors: The current state-of-the-art and recommended applications in biotechnology. *Appl. Microbiol. Biotech.* 86(1): 41–49.

Eibl, R., Rutschmann, K., Lisica, L., and Eibl, D. 2003. Kosten reduzieren durch Einwegbioreaktoren? *BioWorld* 5: 22–23.

Eibl, R., Werner, S., and Eibl, D. 2009. Bag bioreactor based on wave-induced motion: Characteristics and applications. In Eibl, D. and Eibl, R. (eds.), *Disposable Bioreactors, Series: Advances in Biochemical Engineering/Biotechnology*, Vol. 115. Berlin; Heidelberg: Springer, 55–87.

Eibl, R., Werner, S., and Eibl, D. 2009. Bag bioreactor based on wave-induced motion: characteristics and applications. In: Eibl, R. and Eibl, D. (eds.), *Disposable Bioreactors. Series Adv Biochem Eng Biotechnol* 115. Springer, Heidelberg,55–87.

Eibl, R., Werner, S., and Eibl, D. 2009. Disposable bioreactors for plant liquid cultures at litre-scale: Review. *Eng Life Sci* 9: 156–164.

Eibl, D. and Eibl, R. (eds.). 2009. *Disposable Bioreactors, Series: Advances in Biochemical Engineering/Biotechnology*, Vol. 115. Berlin, Heidelberg: Springer.

EMEA/CPMP. 2008. Mark for Guidance on Virus Validation Studies: The Design, Contribution and Interpretation of Studies Validating the Inactivation and Removal of Viruses. Available: http:// www.emea.europa.eu/pdfs/human/bwp/026895en.pdf.

EMEA/CPMP. Mark for Guidance on Virus Validation Studies: The Design, Contribution and Interpretation of Studies Validating the Inactivation and Removal of Viruses. Available: http:// www.emea.europa.eu/pdfs/human/bwp/026895en.pdf. 2008.

EPA 2007. Computational Toxicology Program. Distributed Structure-Searchable Toxicity (DSSTox) Database Network. U.S. Environmental Protection Agency: Washington, DC; http://www.epa.gov/ncct/dsstox.

EPA. 2009. Guide for Industrial Waste Management. Available: www.epa.gov/epawaste/nonhaz/industrial/guide.

Equipment Construction. 2006. Code of Federal Regulations, Food and Drugs Title 21, Part 211.65. U.S. Government Printing Office, Washington, DC; http://www.accessdata.fda.gov/scripts/cdrh/cfdocs/cfcfr/CFRSearch.cfm?CFRPart=211.

Etzel, M. and Riordan, W. 2006. Membrane chromatography: Analysis of breakthrough curves and viral clearance. In Shukla, A., Etzel, M., and Gadam, S. (eds.), *Process Scale Bioseparations for the Biopharmaceutical Industry*. Taylor & Francis, Boca Raton, FL, 277–296.

EU. 1999. Directive 1999/31/EC; criteria and procedures for the acceptance of waste to landfills. *Off. J. Eur. Union* 182: 1–19.

EU. 2000. Directive 2000/76/EC; incineration of waste. *Off J. Eur. Union* 332: 91–111.

EU. 2003. Directive 2002/96/EC; waste electrical and electronic equipment. *Off. J. Eur. Union* 37: 24–38.

EUDRALEX Volume 4: Good Manufacturing Practices, Medicinal Products for Human and Veterinary Use. European Commission: Brussels, Belgium; 1998 http://www.ec.europa.eu/enterprise/pharmaceuticals/eudralex/homev4.htm.

European Commission (Enterprise Directorate General). 2006. EMEA Guideline on Virus Safety Evaluation of Biotechnological Investigational Medicinal Products. London.

European Commission Enterprise Directorate General. 2001. Working party on control of medicines and inspections. EU Guide to Good Manufacturing Practice, Volume 4, Annex 15, July 2001, Cleaning Validation.

European Commission. 1992. Eudralex Vol. 4, Annex 12: Use of ionising radiation in the manufacture of medicinal products. Brussels, Belgium

European Commission. 1992. Eudralex Vol. 4, Annex 14: Manufacture of products derived form human blood or human plasma. Brussels, Belgium.

European Commission. 1992. Eudralex Vol. 4, Annex 2: Manufacture of biological medicinal products for human use. Brussels, Belgium.

European Commission. 2001. Eudralex Vol. 4, Annex 15: Qualification and validation. Brussels, Belgium.

European Commission. 2005. Eudralex Vol. 4, EU Guidelines to Good Manufacturing Practice, part II: Basic requirements for active substances used as starting materials. Brussels, Belgium.

European Commission. 2008. Eudralex Vol. 4, Annex 1: Manufacture of sterile medicinal products. Brussels, Belgium.

European Commission. 2008. Eudralex Vol. 4, EU Guidelines to Good Manufacturing Practice, part I: Basic requirements for medicinal products. Brussels, Belgium.

European Directorate for the Quality of Medicines & Healthcare. 2009. *European Pharmacopeia*, Vol. 6.5, 6th ed. Strasbourg, France: Council of Europe.

European Medicines Evaluation Agency (EMEA). 2005. Guideline on Plastic Immediate Packaging Materials. CHMP/CVMP 205/04.

European Parliament and the Council of the European Union on Packaging and Packaging Waste. 1994. European Parliament and Council Directive 94/62/EC on Packaging and Packaging Waste.; http://www.eur-lex.europa.eu/LexUriServ/LexUriServ.do?uri=CELEX:31994L0062: EN:HTML.

Extractables and Leachables Subcommittee of the Bio-Process Systems Alliance. 2008. Recommendations for extractables and leachables testing. *BioProcess Int.* 6(5): S28–S39.

Fahrner, R.L., Iyver, H.V., and Blank, G.S. 1999. The optimal flow rate and column length for maximum production rate of protein A chromatography. *Bioprocess Eng.* 21: 287–292.

Fahrner, R.L., Knudsen, H.L., Basey, C.D., Galan, W., and Feuerhelm, D. 2001. Industrial purification of pharmaceutical antibodies: Development, operation, and validation of chromatography processes. *Biotechnol. Genet. Eng. Rev.* 18: 301–327.

Falch, F.A. and Heden, C.G. 1963. Disposable shaker flasks. *Biotechnol. Bioeng.* 5: 211–220.

Falkenberg, F.W. 1998. Production of monoclonal antibodies in the miniPerm bioreactor: Comparison with other hybridoma culture methods. *Res. Immunol.* 6: 560–570.

Farid, S.S., Washbrook, J., and Titchener-Hooker, N.J. 2005. Decision-support tool for assessing biomanufacturing strategies under uncertainty: Stainless steel versus disposable equipment for clinical trial material preparation. *Biotechnol. Prog.* 21 (2): 486–497.

Farid, S.S. 2007. Process economics of industrial monoclonal use systems, part 2. *BioProcess Int.* (Suppl. 3): 51–56.

Farid, S.S. 2009. Process economic drivers in industrial monoclonal antibody manufacture. In Gottschalk, U. (ed.), *Process Scale Purification of Antibodies*. New York: Wiley, 239–262.

Farshid, M., Taffs, R.E., Scott, D., Asher, D.M., and Brorson, K. 2005. The clearance of viruses and transmissible spongiform encephalopathy agents from biologicals. *Curr. Opin. Biotechnol.* 16: 561–567.

FDA. 1987. Guideline on Sterile Drug Products Produced by Aseptic Processing. Rockville, MD: FDA.

FDA. 2004. Equipment cleaning and maintenance. Code of Federal Regulations (CFR) Part 211.67 Title 21 Rev. FDA, Rockville, MD, May 25.

Fenge, C. and Lüllau, E. 2006. Cell culture bioreactors. In Ozturk, S.S. and Hu, W.S. (eds.), *Cell Culture Technology for Pharmaceutical and Cell-Based Therapies*. New York: CRC Press, 155–224.

Fermentation—History—Ferments, Living, Cell, Debate, Result, and Cells http://science.jrank.org/pages/2675/Fermentation-History.html#ixzz1IeED0utS.

Fichtner, S., Giese, U., Phal, I., and Reif, O.W. 2006. Determination of "extractables" on polymer materials by means of HPLC-MS. *PDA J. Pharm. Sci. Technol.* 60: 291–301.

Forster, R. and Ishikawa, M. 1999. The methodologies for impact assessment of plastic waste management options—How to handle economic and ecological impacts? Proceedings R99. Geneva.

Foulon, A. et al. 2008. Using disposables in an antibody production process. *BioProcess Int.* 6(6): 12–18.

Foulon, A., Trach, F., Pralong, A., Proctor, M., and Lim, J. 2008. Using disposables in an antibody production process: A cost-effectiveness study of technology transfer between two production sites. *BioProcess Int.* 6 (Suppl 3): 12–18.

Fox, S. 2005. Disposable processing: The impact of disposable bioreactors on the CMO industry. *Contract Pharma.* 6: 62–74.

Fraud, N., Kuczewski, M., Zarbis-Papastoitsis, G., and Hirai, M. 2009. Hydrophobic membrane adsorbers for large-scale downstream processing. *BioPharm. Int.* 23: 24–27.

Fries, S., Glazomitsky, K., Woods, A., Forrest, G., Hsu, A., Olewinski, R., Robinson, D., and Chartrain, M. 2005. Evaluation of disposable bioreactors. *BioProcess Int.* 3(Suppl 6): 36–44.

Fuller, M. and Pora, H. 2008. Introducing disposable systems into biomanufacturing: A CMO case study. *BioProcess Int.* 6(10): 30–36.

Galliher, P. 2008. Achieving high-efficiency production with microbial technology in a single-use bioreactor platform. *BioProcess Int.* 11: 60–65.

Ganzlin, M., Marose, S., Lu, X., Hitzmann, B., Scheper, T., and Rinas, U. 2007. In situ multi-wavelength fluorescence spectroscopy as effective tool to simultaneously monitor spore germination, metabolic activity and quantitative protein production in recombinant *Aspergillus niger* fed-batch cultures. *J. Biotechnol.* 132 (4): 461–468.

GE Healthcare. 2006. CHO cell supernatant concentration with Kvick Lab cassettes. Application note 11-0013-62.

GE Healthcare. 2007. Purification of a monoclonal antibody using ReadyToProcess columns. Application Note 28–9198–56 AA.

GE Healthcare. 2008. Rapid production of clinical grade T lymphocytes in the Wave Bioreactor. Available: http://www.5.gelifesciences.com/aptrix/upp0091.nsf./Content/F7AD616DACC22171C125747400812B51/$file/28933149AA.pdf.

GE Healthcare. 2009. A flexible antibody purification process based on ReadyToProcess products. Application note 28-9403-48 AA.

GE Healthcare. 2009. High-throughput screening and optimi-zation of a multi modal polishing step in a monoclonal antibody purification process. Application note 28-9509-60.

GE Healthcare. 2009. High-throughput screening and optimization of a Protein A capture step in a monoclonal antibody purification process. Application note 28-9468-58.

GE Healthcare. 2009. Scale-up of a downstream monoclonal antibody purification process using HiScreen and AxiChrom columns. Application note 28-9409-49.

GE Healthcare. 2010. Hot Lips Tube Sealer—Product description. Available: http://www.gelifesciences.com/aptrix/upp01077.nsf/Content/wave_bioreactor_home-wave_fluid_transfer-hot_lips_tube_sealer.

GE Healthcare. 2010. ReadyMate DAC—Product description. Available: http://www5.gelifesciences.com/aptrix/upp00919.nsf/Content/692F8252BA8B1477C125763C00827AEI/$file/28937902+AC+.pdf.

GE Healthcare. 2010. Sterile Tube Fuser—Technical information. Available: http://www5.gelifesciences.com/aptrix/upp01077.nsf/Content/Products?OpenDocument&parentid=986919&moduleid=167710&zone=.

GE Healthcare. 2010. Wave Mixer Concept. Available: http://wwwI.gelifesciences.com/aptrix/upp01077.nsf/Content/wave_bioreactor_home-how_it_works-how_it_works_wave_mixer.

GE. 2009. NPC-100 Pressure Sensor. Available: http://www.gesensing.com/products/npc_100_series.htm?bc=bc_indust+bc_med_fluid.

Gebauer, A., Scheper, T., and Schugerl, K. 1987. Penicillin acylase production by *E. coli. Bioprocess Eng.* 2 (2): 55–58.

Gebauer, K., Thommes, J., and Kula, M. 1997. Plasma protein fractionation with advanced membrane adsorbents. *Biotechnol. Bioeng.* 54: 181–189.

Gebauer, K.H., Thommes, J., and Kula, M. 1997. Breakthrough performance of high capacity membrane adsorbers in protein chromatography. *Chew. Eng. Sci.* 52: 405–419.

Genzel, Y., Olmer, R.M., Schaefer, B., and Reichl, U. 2006. Wave microcarrier cultivation of MDCK cells for influenza virus production in serum containing and serum-free media. *Vaccine* 24: 6074–6087.

Georgiev, M.I., Weber, J., and Maciuk, A. 2009. Bioprocessing of plant cell cultures for mass propagation of targeted compounds. *Appl. Microbiol. Biotechnol.* 83: 809–823.

Ghosh, R. 2004. Protein separation using membrane chromatography: Opportunities and challenges. *J. Chromatogr. A* 952: 13–27.

Girard, L.S., Fabis, M.J., Bastin, M., Courtois, D., Pétiard, V., and Koprowski, H. 2006. Expression of a human anti-rabies virus monoclonal antibody in tobacco cell culture. *BBRC* 345: 602–607.

Glaser, V. 2005. Disposable bioreactors become standard fare. *Genet. Eng. Biotechnol. News* 25 (14): 80–81.

Glaser, V. 2009. Bioreactor and fermentor trends. *Genet. Eng. Biotechnol. News.* Available: http://www.genengnews.com/issues/item.aspx.

Gold, L.S. Carcinogenic Potency Database (CPDB). University of California, Berkeley, CA; http://www.potency.berkeley.edu/cpdb.html. 2007.

Goldstein, A., Loesch, J., Mazzarella, K., Matthews, T., Luchsinger, G., and Javier, D.S. 2009. Freeze Bulk Bags: A Case Study in Disposables Implementation: Genentech's Evaluation of Single-Use Technologies for Bulk Freeze-Thaw, Storage, and Transportation. Available: http://biopharminternational.findpharma.com/biopharm/Disposables+Articles/Freeze-Bulk-Bags-A-Case-Study-in-Disposables-lmple/ArticleStandard/Article/detail/637583.

Gorter, A., van de Griend, R.J., van Eendenburg, J.D., Haasnot, W.H., and Fleuren, G.J. 1993. Production of bi-specific monoclonal antibodies in a hollow-fibre bioreactor. *J. Immunol. Methods* 161: 145–150.

Gottschaik, U. 2005. Downstream processing of monoclonal antibodies: From high dilution to high purity. *BioPharm. Int.* 19: 42–58.

Gottschaik, U. 2008. Bioseparation in antibody manufacturing: The good, the bad and the ugly. *Biotechnol. Prog.* 24: 496–503.

Gottschalk, U., Fischer-Fruehholz, S., and Reif, O.W. 2004. Membrane adsorbers: A cutting-edge process technology at the threshold. *BioProcess Int.* 2: 56–65.

Gottschalk, U. 2005. New and unknown challenges facing biomanufacturing. *BioPharm. Int.* 19: 24–28.

Gottschalk, U. 2005. Thirty Years of Monoclonal Antibodies: oCSingle_Use_Disposables.pdf. A Long Way to Pharmaceutical and Commercial Success Modern Pharmaceuticals. Weinheim: Wiley-VCH Verlag GmbH & Co. KGaA.

Gottschalk, U. 2006. The renaissance of protein purification. *BioPharm. Int.* 6 (Suppl. 6): 8–9.

Gottschalk, U. 2007. New standards in virus and contaminant clearance. *BioPharm. Int.* 10: 5.

Gottschalk, U. 2007. The renaissance of protein purification. *BioPharm. Int.* 10: 41–42.

Gottschalk, U. 2009. Disposables in downstream processing. In Eibl, R. and Eibl, D. (eds.), *Disposable Bioreactors, Series: Advances in Biochemical Engineering/Biotechnology*, Vol. 115. Berlin; Heidelberg: Springer, 172–183.

Greb, E. 2009. The debate over preuse filter-integrity testing. PharmTech. Available: http://pharmtech.findpharma.com/pharmtech/Article/The-Debate-over-Preuse-Filter-lntegrity-Testing/ArticleStandard/Article/detail/612612.

Groton Biosystems. 2009. 24/7 online reactor autosampling. Available: http://www.grotonbiosystems.com/pressroom/press_docs/MKT-046_ProductBulletin_SampleProbe.pdf.

Guide for Industrial Waste Management. 2008. US Environmental Protection Agency: Washington, DC; http://www.epa.gov/epawaste/nonhaz/industrial/guide/index.htm.

Guide for Industrial Waste Management. US Environmental Protection Agency; http://www.epa.gov/epawaste/nonhaz/industrial/guide/index.htm.

Gupta, A. and Rao, G. 2003. A study of oxygen transfer in shake flasks using a non-invasive oxygen sensor. *Biotechnol. Bioeng.* 84: 351–358.

Guth, U., Gerlach, F., Decker, M., Oelßner, W., and Vonau, W. 2009. Solid-state reference electrodes for potentiometric sensors. *J. Solid State Electrochem.* 13 (1): 27–39.

Hagedorn, A., Levadoux, W., Groleau, D., and Tartakovsky, B. 2004. Evaluation of multiwavelength culture fluorescence for monitoring the aroma compound 4-hydroxy-2(or 5)-ethyl-5(or 2)-methyl-3(2H)-furanone (HEMF) production. *Biotechnol. Prog.* 20 (1) 361–367.

Haldankar, R., Li, D., Saremi, Z., Baikalov, C., and Deshpande, R. 2006. Serum-free suspension large-scale transient transfection of CHO cells in wave bioreactors. *Mol. Biotechnol.* 34: 191–199.

Hami, L.S., Ghana, H., Yuan, V., and Craig, S. 2003. Comparison of a static process and a bioreactor-based process for the GMP manufacture of autologous Xcellerated T cells for clinical trials. *BioProcessing J.* 2: 1–10.

Hami, L.S., Green, C., Leshinsky, N., Markham, E., Miller, K., and Craig, S. 2004. GMP production of Xcellerated T cells for the treatment of patients with CLL. *Cytotherapy* 6: 554–562.

Hamid Mollah, A. 2008. Cleaning validation for bio-pharmaceutical manufacturing at Genentech. *BioPharm. Int.* 2: 36–41.

Hamis, L.S., Green, C., Leshinsky, N., Markham, E., Miller, K., and Craig, S. 2004. GMP production and testing of Xcellerated T cells for the treatment of patients with CLL. *Cytotherapy* 6: 554–562.

Hantelmann, K., Kollecker, A., Hull, D., Hitzmann, B., and Scheper, T. 2006. Two-dimensional fluorescence spectroscopy: A novel approach for controlling fed-batch cultivations. *J. Biotechnol.* 121 (3): 410–417.

Harris, M. 2009. The Billion-Plus Blockbusters: The Top 25 Biotech Drugs. BioWorld (http://www.bioworld.com/servlet/com.accumedia.web.Dispatcher?next=bioWorldHeadlines_article&forceid=51907.

Haughney, H. and Hutchinson, J. 2004. A disposable option for bovine serum filtration and packaging. *BioProcess Int.* Suppl. 4 (9): 2–5.

Haughney, H. and Hutchinson, J. 2004. Single-use systems reduce production time-lines. *Gen. Eng. News* 24(8); www.hyclone.com/pdf/bp_single_use24gen8.pdf.

Heath, C. and Kiss, R. 2007. Cell culture process development: Advances in process engineering. *Biotechnol. Prog.* 23: 46–51.

Heinzle, E., Biwer, A.P., and Cooney, C.L. 2006. Monoclonal antibodies. In Heinzle, E., Biwer, A.P., and Cooney, C.L. (eds.), *Development of Sustainable Bioprocesses-Modelling and Assessment.* Chichester, UK: John Wiley & Sons, 241–260.

Heller, A. and Feldmann, B. 2008. Electrochemical glucose sensors and their applications in diabetes management. *Chem. Rev.*108: 2482–2505.

Hemmerich, K.J. 2000. Polymer Materials Selection for Radiation-Sterilized Products. MDDI; http://www.devicelink.com/mddi/archive/00/02/006.html.

Henning, B. and Rautenberg, J. 2006. Process monitoring using ultrasonic sensor systems. *Ultrasonics* 44: e1395–e1399.

Henzler, H.J.. 2000. Particle stress in bioreactors. In K. Schügerl, G. Kretzmer (eds.), *Influence of Stress on Cell Growth and Product Formation, Series: Advances in Biochemical Engineering/Biotechnology,* Vol. 67. Berlin; Heidelberg: Springer, 38–82.

Hess, S., Baier, U., Lettenbauer, C., and Hafner, D. 2002. A new application for the wave bioreactor 20: Cultivation of Erynia neoaphidis, a mycel producing fungus. IOBC Meeting "Insect pathogens and insect parasitic nematodes." Birmingham, UK.

High cell density bioreactor offload clarification using Sartoclear® P Depth Filters. Application Note 5, Sartorius. Sartorius Stedim Biotech 2008.

Higuchi, A., Kyokon, M., Murayama, S., Yokogi, M., Hirasaki, T., and Manabe, S.I. 2004. Effect of aggregated protein sizes on the flux of protein solution through microporous membranes. *J. Memb. Sci.* 236: 137–144.

Hilmer, J.-M. and Scheper, T. 1996. A new version of an in situ sampling system for bioprocess analysis. *Acta Biotechnol.* 16 (2–3): 185–192.

Hirschy, O., Schmid, T., Grunder, J.M., Andermatt, M., Bollhalder, F., and Sievers, M. 2001. Wave reactor and the liquid culture of the entomopathogenic nematode Steinerma feltiae. In Griffin, C.T., Burnell, A.M., Downes, M.J., and Mulder, R. (eds.), *Developments in Entomopathogenic Nematode/Bacterial Research.* DG XII, COST 819, Brussels, Luxembourg.

Hitchcock, T. 2009. Production of recombinant whole-cell vaccines with disposable manufacturing systems. *BioProcess Int.* 5: 36–43.

Hitzmann, B., Broxtermann, O., Cha, Y.L., Sobieh, O., Stark, E., and Scheper, T. 2000. The control of glucose concentration during yeast fed-batch cultivation using a fast measurement complemented by an extended Kalman filter. *Bioprocess Eng.* 23 (4): 337–341.

Hopkinson, J. 1985. Hollow fibre cell culture systems for economical cell-product manufacturing. *BioTechnology* 3: 225–230.

Horowitz, B., Lazo, A., Grossberg, H., Page, G., Lippin, A., and Swan, G. 1998. Virus inactivation by solvent/detergent treatment and the manufacture of SD-plasma. *Vox Song.* (Suppl. I): 203–206.

Hou, K. and Mandaro, R. 1986. Bioseparation by ion exchange cartridge chromatography. *BioTechniques* 4: 358–366.

Houtzager, E., van der Linden, R., de Roo, G., Huurman, S., Priem, P., and Sijmons, P.C. 2005. Linear scale-up of cell cultures. The next level in disposable bioreactor design. *BioProcess Int.* 6: 60–66.

Hsiao, T.Y., Bacani, F.T., Carvalho, E.B., and Curtis, W.R. 1999. Development of a low capital investment reactor system: Application for plant cell suspension culture. *Biotechnol Prog* 15: 114–122.

Hundt, B., Best, C., Schlawin, N., Kassner, H., Genzel, Y., and Reichl, U. (2007). Establishment of a mink enteritis vaccine production process in stirred-tank reactor and Wave®Bioreactor microcarrier cultures in 1–10 L scale. *Vaccine* 25: 3987–3995.

Hynetics Co. 2010. Description of the HyNetics disposable mixing system. Available: http://www.hynetics.com/default/WhatWeOffer.htm.

HyNetics. 2010. Available: http://www.hynetics.com.

ICH Q7: Good Manufacturing Practice Guide for Active Pharmaceutical Ingredients. Federal Register 66(186): 49028–49029; 2000, http://www.ich.org/LOB/media/MEDIA433.pdf.

Immelmann, A., Kellings, K., Stamm, O., and Tarrach, K. 2005. Validation and quality procedures for virus and prion removal in biopharmaceuticals. *BioProcess Int.* 3: 38–45.

Incardona, N.L., Tuech, J.K., and Murti, G. 1985. Irreversible binding of phage phi XI74 to cell-bound lipopolysaccharide receptors and release of virus-receptor complexes. *Biochemistry* 24: 6439–6446.

International Conference on Harmonization. 1995. Q5B: Quality of biotechnological products: Analysis of the expression construct in cells used for production of r-DNA derived protein products. Geneva, Switzerland.

International Conference on Harmonization. 1995. Q5C: Quality of biotechnological products: Stability testing of biotechnological/biological products. Geneva, Switzerland.

International Conference on Harmonization. 1997. Q5D: Derivation and characterisation of cell substrates used for production of.

International Conference on Harmonization. 1997. S6: Preclinical safety evaluation of biotechnology-derived pharmaceuticals. Geneva, Switzerland.

International Conference on Harmonization. 1999. Q5A: Viral safety evaluation of biotechnology products derived from cell lines of human or animal origin Q5A. Geneva, Switzerland.

International Conference on Harmonization. 2000. Q7: Good manufacturing practice guide for active pharmaceutical ingredients. Geneva, Switzerland.

International Conference on Harmonization. 2002. Q5E: Comparability of biotechnological/biological products subject to changes in their manufacturing process. Geneva, Switzerland.

International Conference on Harmonization. 2005. Q9: Quality risk management Geneva, Switzerland.

International Conference on Harmonization. 2006. Q3A: Impurities in new drug substances. Geneva, Switzerland.

International Conference on Harmonization. 2006. Q3B: Impurities in new drug products. Geneva, Switzerland.

International Conference on Harmonization. 2007. Q3C: Impurities: Guideline for residual solvents. Geneva, Switzerland.

International Conference on Harmonization. 2008. Q10: Pharmaceutical quality management system. Geneva, Switzerland.

International Conference on Harmonization. 2008. Q8: Pharmaceutical development. Geneva, Switzerland.

International Organization for Standardization. 2006. Sterilization of health care products—Radiation. ISO 11137, ISO's Central Secretariat. Geneva, Switzerland.

International Organization for Standardization. 2006. Sterilization of health care products—Biological indicators. ISO 11138, ISO's Central Secretariat. Geneva, Switzerland.

International Organization for Standardization. 2006. Sterilization of health care products—Vocabulary. ISO 11139, ISO's Central Secretariat. Geneva, Switzerland.

International Organization for Standardization. 2008. Sterilization of health care products—Ethylene oxide. ISO 11135, ISO's Central Secretariat. Geneva, Switzerland.

International Organization for Standardization. 2009. Sterilization of health care products—General requirements for characterization of sterilizing agent and the development, validation and routine control of sterilization process for medical devices. ISO 14937, ISO's Central Secretariat. Geneva, Switzerland.

International Organization for Standardization. 2009. Sterilization of health care products—biological indicators—guidance for the selection, use and interpretation of results. ISO 14161, ISO's Central Secretariat. Geneva, Switzerland.

Internationally Conference on Harmonization. 1998. Q5A: Viral safety evaluation of biotechnology product derived from cell lines of humanly or animal origin. Geneva, Switzerland.

Ireland, T., Lutz, H., Siwak, M., and Bolton, G. 2004. Viral filtration of plasma-derived humanly IgG: A case study using VIRESOLVE NFR *BioPharm. Int. I* 1: 38–44.

Isberg, E.A. 2007. Advanced aseptic processing: RABS and isolator operations. *Pharm. Eng.* 27: 18–21.

Ishihara, T. and Kadoya, T. 2007. Accelerated purification process development of monoclonal antibodies for shortening time to clinic design and case study of chromatography processes. *J. Chromatogr. A* 176: 149–156.

Iwaya, M., Iron Mountain, S., Bartok, K., and Denhardt, G. 1973. Mechanism of replication of single stranded PhiXI74 DNA. VII. Circularization of the progeny viral strand. *J. Virol.* 12: 808–818.

Jablonski-Lorin, C., Mellio, V., and Hungerbühler, E. 2003. Stereoselective bioreduction to a chiral building block on a kilogram scale. *Chimia* 57: 574–576.

Jagschies. G. and O'Hara, A. 2007. Debunking downstream bottleneck myth. *Genet. Eng. News* 27: 3.

Jagschies, G. 2009. Flexible manufacturing: Driving monoclonal antibody process economics. *GIT BIOprocessing* 1: 30–31.

Jain, E. and Kumar, A. 2008. Upstream processes in antibody production: Evaluation of critical parameters. *Biotechnol. Adv.* 26: 46–72.

Jenke, D., Story, J., and Lalani, R. 2006. Extractables/leachables from plastic tubing used in product manufacturing. *Int. J. Pharm.* 315: 75–92.

Jenke, D. 2007. An extractables/leachables strategy facilitated by collaboration between drug product vendors and plastic material/system suppliers. *PDA J. Pharm. Sci. Technol.* 61: 17–23.

Jenke, D. 2007. Evaluation of the chemical compatibility of plastic contact materials and pharmaceutical products; safety consider-ations related to extractables and leachables. *J. Pharm. Sci.* 96: 2566–2581.

Jia, Q., Li, H., Hui, M., Hui, N., Joudi, A., Rishton, G., Bao, L., Shi, M., Zhang, X., Luanfeng, L., Xu, J., and Leng, G. 2008. A bioreactor system based on a novel oxygen transfer method. *BioProcess Int.* 6: 66–78.

Joeris, K., Frerichs, J.G., Konstantinov, K., and Scheper, T. 2002. In-situ microscopy: Online process monitoring of mammalian cell cultures. *Cytotechnology* 38 (1–2): 129–134.

Joeris, K. and Scheper, T. 2003. Visualizing transport processes at liquid-liquid interfaces—The application of laser-induced fluorescence. *J. Colloid Interface Sci.* 267 (2): 369–376.

Jones C. and Dent, A. 2007. Integration of generous scale chromatography with nanofiltration for in ovine polyclonal product. 2nd European Downstream Technology Forum, Goettingen, Germany.

Jones, S.C.B. and Smith, M.P. 2004. Evaluation of an alkali stable Protein A matrix versus Protein A Sepharose Fast Flow and considerations on process scale-up to 20,000 L. 3rd International Symposium on Downstream Processing of Genetically Engineered Antibodies and Related Molecules, Nice, France, October.

Joo, S. and Brown, R.B. 2008. Chemical sensors with integrated electronics. *Chem. Rev.* 108 (2): 638–651.

Jornitz, M.W. and Meltzer, T.H. 2004. *Filtration Handbook Liquids.* River Grove, IL: PDA, DHI LLC.

Jornitz, M.W. and Meltzer, T.H. 2006. *Pharmaceutical Filtration—The Management of Organism Removal.* River Grove, IL: PDA, DHI LLC.

Jornitz, M.W. and Meltzer, T.H. (eds.). 2008. Filtration and purification in the biopharmaceutical industry. In SwarbrickJ (ed.), *Drugs and Pharmaceutical Sciences*, 2nd ed., Vol. 174. New York: Informa Healthcare.

Kallenbach, N.O., Cornelius, P.A., Negus, D., Montgomerie, D., and Englander, S. 1989. Inactivation of viruses by ultraviolet light. *Curr. Stud. Hematol. Blood Transfus.* 56: 70–82.

Karlsson, E., Ryden, L., and Brewer, J. 1989. Ion exchange chromatography. In Janson, J.C. and Ryden, L. (eds.), *Protein Purification: Principles, High-Resolution Methods and Applications.* New York: VCH, 107–115.

Kato, Y., Peter, C.P., Akgün, A., and Büchs, J. 2004. Power consumption and heat transfer resistance in large rotary shaking vessels. *Biochem. Eng. J.* 21: 83–91.

Kelley, B. 2009. Industrialization of mAb production technology. The bioprocessing industry at a crossroads. *MAbs* 1(5): 443–452.

Kelley, K. 2007. Very large scale monoclonal antibody purification: The case for conventional unit operations. *Biotechnol. Prog.* 23: 995–1008.

Kemeny, M.M., Cooke, V., Melester, T.S., Halperin, I.C., Burchell, A.R., Yee, J.P., and Mills, C.B. 1993. Splenectomy in patients with AIDS and AIDS-related complex. *AIDS* 7 (8): 1063–1067.

Kemplen, R., Preissmann, A., and Berthold, W. 1995. Assessment of a disc stack centrifuge for use in mammalian cell separation. *Biotechnol. Bioeng.* 46: 132–138.

Kermis, H.R., Kostov, Y., and Rao, G. 2003. Rapid method for the preparation of a robust optical pH sensor. *Analyst* 128 (9): 1181–1186.

Khanna, V. 2008. ISFET (ion-sensitive field-effect transistor)-based enzymatic biosensors for clinical diagnostics and their signal conditioning instrumentation. *IETE J. Res.* 54 (3): 193–202.

Khanna, V.K., Ahmad, S., Jain, Y.K., Jayalakshmi, M., Vanaja, S., Madhavendra, S.S., Manorama, S.V., Kantam, M.L. 2007. Development of potassium-selective ion-sensitive field-effect transistor (ISFET) by depositing ionophoric crown ether membrane on the gate dielectric, and its application to the determination of K+-ion concentrations in blood serum. *IJEMS* 14 (2): 112–118.

Khanna, V.K., Kumar, A., Jain, Y.K., and Ahmad, S. 2006. Design and development of a novel high-transconductance pH-ISFET (ion-sensitive field-effect transistor)-based glucose biosensor. *Int. J. Electron.* 93 (2): 81–96.

Kinney, S.D., Phillips, C.W., and Lin, K.J. 2007. Thermoplastic tubing, welders and sealers. *BioProcess Int.* 5 (4): 52–61.

Kleenpak data sheet biopharm_34125. Pall Corporation 2009.

Klimant, I., Kuhl, M., Glud, R.N., and Hoist, G. 1997. Optical measurement of oxygen and temperature in microscale: Strategies and biological applications. *Sens. Actuators B Chem.* 38 (1–3): 29–37.

Knazek, R.A., Gullino, P.M., Kohler, P.O., and Dedrick, R.L. 1972. Ceil culture on artificial capillaries: An approach to tissue growth in vitro. *Science* 178: 65–67.

Knevelman, C., Hearle, D.C., Osman, J.J., Khan, M., Dean, M., Smith, M., and Aiyedebinu Cheung, K. 2002. Characterization and Operation of a Disposable Bioreactor as a Replacement for Conventional Steam-In-Place Inoculum Bioreactors for Mammalian Cell Culture Processes. Available: http://www.5.gelifesciences.com/aptrix/upp01077.nsf/Content/wave_bioreactor_home-wave_literature_WindowsInternetExplorer.

Knudsen, H.L., Fahrner, R.L., Xu, Y., Norling, L.A., and Blank, G.S. 2001. Membrane ion-exchange chromatography for processscale antibody purification. *Chromatogr. A* 907: 145–154.

Knuttel, T., Hartmann, T., Meyer, H., and Scheper, T. 2001. On-line monitoring of a quasi-enantiomeric reaction with two coumarin substrates via 2D-fluorescence spectroscopy. *Enzyme Microb. Technol.* 29 (2–3): 150–159.

Knuttel, T., Meyer, H., and Scheper, T. 2005. Application of 2D-fluorescence spectroscopy for on-line monitoring of pseudoenantiomeric transformations in supercritical carbon dioxide systems. *Anal. Chem.* 77 (19): 6184–6189.

Knuttel, T., Meyer, H., and Scheper, T. 2006. The application of two-dimensional fluorescence spectroscopy for the on-line evaluation of modified enzymatic enantioselectivities in organic solvents by forming substrate salts. *Enzyme Microb. Technol.* 39 (4): 607–611.

Kohls, O. and Scheper, T. 2000. Setup of a fiber optical oxygen multisensor-system and its applications in biotechnology. *Sens. Actuators B Chem.* 70 (1–3): 121–130.

Koneke, R., Comte, A., Jurgens, H., Kohls, O., Lam, H., and Scheper, T. 1998. Faseroptische Sauerstoffsensoren für Biotechnologie, Umwelt- und Lebensmitteltechnik. *Chemie Ingenieur Technik* 70 (12): 1611–1617.

Kornmann, H., Valentinotti, S., Duboc, P., Marison, I., and von Stockar, U. 2004. Monitoring and control of Gluconacetobacter xylinus fed-batch cultures using in situ MID-IR spectroscopy. *J. Biotechnol.* 113: 231–245.

Kranjac, D. 2004. Validation of bioreactors: Disposable vs. reusable. *BioProcess Int. Industry Yearbook*: 86.

Kraume, M. (ed.). 2003. *Mischen und Rühren-Grundlagen und moderne Verfahren.* Weinheim: Wiley-VCH Verlag GmbH & Co. KGaA.

Kresta, S.M. and Brodkey, R.S. 2004. Turbulence in mixing applications. In Paul, E.L., Atiemo-Obeng, V.A., and Kresta, S.M. (eds.), *Handbook of Industrial Mixing-Science and Practice.* Hoboken, NJ: John Wiley & Sons, 19–88.

Kubota, N., Konno, Y., Miura, S., Saito, K., Sugita, K., Watanabe, K., and Sugo, T. 1996. Comparison of two convection-aided protein adsorption methods using porous membranes and perfusion beads. *Biotechnol. Prog.* 12: 729–876.

Kullick, T., Bock, U., Schubert, J., Scheper, T., and Schugerl, K. 1995. Application of enzyme-field effect transistor sensor arrays as detectors in a flow-injection analysis system for simultaneous monitoring of medium components. Part II. Monitoring of cultivation processes. *Anal. Chim. Acta* 300 (1–3): 25–31.

Kullick, T., Quack, R., Röhrkasten, C., Pekeler, T., Scheper, T., and Schügerl, K. 1995. Pbs-field-effect-transistor for heavy metal concentration monitoring. *Chem. Eng. Technol.* 18 (4): 225–228.

Kybal, J. and Sikyta, B. 1985. A device for cultivation of plant and animal cells. *Biotechnol. Lett* 7: 467–47.

Kybal, J. and Vlcek, V. 1976. A simple device for stationary cultivation of microorganisms. *Biotechnol. Bioeng.* 18: 1713–1718.

Lam, P. and Sane, S. 2007. Design and Testing of a Prototype Large-Scale Bag Freeze-Thaw System: The Development of a Large-Scale Bag Freeze-Thaw System will have Many Benefits for The Biopharmaceutical Industry. Available: http://biopharminternational.findpharma.com/biopharm/Disposables/Design-and-Testing-of-a-Prototype-Large-Scale-Bag-/ArticleStandard/Article/detail/473322.

Lamproye, A. 2006. Viral clearance by nanofiltration, strategies for successful validation studies. 1st European Downstream Forum, Goettingen, Germany.

Landon, R. and Baloda, S. 2005. Disposable technology: Validation of a novel disposable connector for sterile fluid transfer. *BioProcess Int. Industry Yearbook*: 88.

Langer, E. 2008. Fifth Annual Report and Survey of Biopharmaceutical Manufacturing Capacity and Production. Rockville, MD: BioPlan Associates Inc..

Langer, E. 2009. Sixth Annual Report and Survey of Biopharmaceutical Manufacturing Capacity and Production. Rockville, MD: BioPlan Associates.

Langer, E.S. and Price, B.J. 2007. Biopharmaceutical disposables as a disruptive future technology. *BioPharm. Int.* 20 (6): 48–56.

Langer, E.S. and Ranck, J. 2005. The ROI case: Economic justification for disposables in biopharmaceutical manufacturing. *BioProcess Int.* 27 (Suppl. 1): 46–50.

Langer, E.S. 2009. Trends in single-use bioproduction. *BioProcess* 7, March 2009.

Lawrence, N.S., Pagels, M., Hackett, S.F.J., McCormack, S., Meredith, A., Jones, T.G.J., and Wildgoose, G.G., Compton, R.G., and Jiang, L. 2007. Triple component carbon epoxy pH probe. *Electroanalysis* 19 (4): 424–428.

Lee, S.K., Suh, E.K., Cho, N.K., Park, H.D., Uneus, L., and Spetz, A.L. 2005. Comparison study of ohmic contacts to 4H-silicon carbide in oxidizing ambient for harsh environment gas sensor applications. *Solid State Electron.* 49 (8): 1297–1301.

Lee, S.Y., Kim, S.H., Kim, V.N., Hwang, J.H., Jin, M., Lee, J., and Kim, S. 1999. Heterologous gene expression in avian cells: Potential as a producer of recombinant proteins. *J. Biomed. Sci.* 6: 8–17.

Lehmann, J., Heidemann, R., Riese, U., Lütkemeyer, D., and Büntemeyer, H. 1992. Der Superspinner—Ein Brutschrankfermenter für die Massenkultur tierischer Zellen. *BioEngineering* 5/6: 112–117.

Leventis, H.C., Streeter, I., Wildgoose, G.G., Lawrence, N.S., Jiang, L., Jones, T.G.J., and Compton, R.G. 2004. Derivatised carbon powder electrodes: Reagentless pH sensors. *Talanta* 63 (4): 1039–1051.

Levine, B. 2007. Making Waves in Cell Therapy: The Wave Bioreactor for the Generation of Adherent and Non-Adherent cells for Clinical Use. Available: http://www.wavebiotech.com/pdf/literature/ISCT_2007_Levine_Final.pdf.

Lewcook, A. 2007. Disposables crucial in future of patient-specific meds. Available: http://www.in-pharmatechnologist.com/news/ng.asp?id=77560.

Li, C.Y., Zhang, X.B., Han, Z.X., Akermark, B., Sun, L.C., Shen, G.L., and Yu, R.Q. 2006. A wide pH range optical sensing system based on a sol-gel encapsulated amino-functionalised corrole. *Analyst* 131 (3): 388–393.

Lim, J.A.C., Sinclair, A., Kim, D.S., and Gottschalk, U. 2007. Economic benefits of single-use membrane chromatography in polishing. A cost of goods model. *BioProcess Int.* 5: 60–64.

Lim, J.A.C., Sinclair, A., Kim, D.S., and Gottschalk, U. 2007. Economic benefits of single use membrane chromatography in polishing. A cost of goods model. *BioProcess Int.* 2: 60–64.

Lim, J.A.C. and Sinclair, A. 2007. Process economy of disposable manufacturing: Process models to minimize upfront investment. *Am. Pharm. Rev.* 10: 114–121.

Limke, T. 2009. Comparability between the Mobius CellReady 3 L bioreactor and 3 L glass bioreactors. *BioProcess Int.* 7: 122–123.

Lindemann, C., Marose, S., Scheper, T., Nielsen, H.O., and Reardon, K.F. 1999. Fluorescence techniques for bioprocess monitoring. In *Encyclopedia of Bioprocess Technology: Fermentation, Biocatalysis, and Bioseparation*, Flickinger, M.C. and Drew, S.W. (eds.), New York: Wiley.

Liu, C.M. and Hong, L.N. 2001. Development of a shaking bio-reactor system for animal cell cultures. *Biochem. Eng. J.* 2: 121–125.

Liu, C.Z., Towler, M.J., Medrano, G., Cramer, C.L., and Weathers, P.J. 2009. Production of mouse interleukin-12 is greater in tobacco hairy roots grown in a mist reactor than in an airlift reactor. *Biotechnol. Bioeng.* 102: 1074–1086.

Liu, P 2005. Strategies for optimizing today's increasing disposable processing environments. *BioProcess Int.* 9: 10–15.

Liu, P. 2005. Strategies for optimizing today's increasing disposable processing environments. *BioProcess Int.* 3(9): S10–S15.

Liu, P. 2005. Strategies for optimizing today's increasingly disposable processing environments. *BioProcess Int.* 3(6, Supplement): 10–15.

Lloyd-Evans, P., Phillips, D.A., Wright, A.C.C., and Williams, R.K. 2007. Disposable process for cGMP manufacture of plasmid DNA. *BioPharm Int.* 1: 18–24.

Lok, M. and Blumenblat, S. 2007. Critical design aspects of single-use systems: Some points to consider for successful implementation. *BioProcess Int.* Suppl. 5 (S): 28–31.

Lorenz, C.M., Wolk, B.M., Quan, C.P., Alcal, E.W., Closely, M., McDonald, D.J., and Matthew, T.C. 2009. The effect of low intensity ultraviolet C light on monoclonal antibodies. *Biotechnol. Prog.* 25: 476–482.

Lorenzelli, L., Margesin, B., Martinoia, S., Tedesco, M.T., and Valle, M. 2003. Bioelectrochemical signal monitoring of in-vitro cultured cells by means of an automated microsystem based on solid state sensor-array. *Biosens. Bioelectron.* 18 (5–6): 621–626.

Lu, J.Z. and Rosenzweig, Z. 2000. Nanoscale fluorescent sensors for intracellular analysis. *Fresenius J. Anal. Chem.* 366 (6–7): 569–575.

Lute, P., Bailey, M., Combs, J., Sukumar, M., and Brorson, K. 2007. Phage passage after extensive processing in small-virus retentive filters. *Biotechnol. Appl. Biochem.* 47: 141–151.

Lytle, C. and Sagripanti, J. 2005. Predicted inactivation of viruses of relevance to biodefense by solar radiation. *J. Virol.* 79: 14244–14252.

Maa, Y.F. and Hsu, C.C. 1999. Performance of sonication and microfluidization for liquid-liquid emulsification. *Pharm. Dev. Technol.* 4 (2): 233–240.

Mach, C.J. and Riedman, D. 2008. Reducing microbial contamination risk in biotherapeutic manufacturing: Validation of sterile connections. *BioProcess Int.* 6(8): 20–26.

Maharbiz, M.M., Holtz, W.J., Howe, R.T., and Keasling, J.D. 2004. Microbioreactor arrays with parametric control for high-throughput experimentation. *Biotechnol. Bioeng.* 86 (4): 485–490.

Maier, U. and Büchs, J. 2001. Characterization of the gas-liquid mass transfer in shaking bioreactors. *Biochem. Eng. J.* 7: 99–106.

Maier, U., Losen, M., and Büchs, J. 2003. Advances in understanding and modeling the gas–liquid mass transfer in shake flasks. *Biochem Eng J.* 17: 155–167.

Mardirosian, D., Guertin, P., Crowell, J., Yetz-Aldape, J., Hall, M., Hodge, G., Jonnalagadda, K., Holmgren, A., and Galliher, P. 2009. Scaling up a CHO-produced hormone-protein fusion product. *BioProcess Int.* 7 (Suppl. 4): 30–35.

Margesin, R., Schneider, M., and Schinner, F. 1995. *Praxis der biotechnologischen Abluftreinigung.* Berlin: Springer.

Marks, D.M.. 2003. Equipment design considerations for large scale cell culture. *Cytotechnology* 42: 21–33.

Marose, S., Lindemann, C., and Scheper, T. 1998. Two-dimensional fluorescence spectroscopy: A new tool for on-line bioprocess monitoring. *Biotechnol. Prog.* 14 (1): 63–74.

Marose, S., Lindemann, C., Ulber, R., and Scheper, T. 1999. Optical sensor systems for bioprocess monitoring. *Trends Biotechnol.* 17 (1): 30–34.

Martin, J. 2004. Case study: Orthogonal Membrane Technologies for Virus and DNA Clearance. PDA International Congress, Rome, Italy.

Marx, U. 1998. Membrane-based cell culture technologies: A scientifically economically satisfactory alternative malignant ascites production for monoclonal antibodies. *Res. Immunol.* 6: 557–559.

Maslowski, J. 2007. Available: http://www.asepticfilling.com/PDC-disposable.htm.

Masser, D. 2008. Advanced Scientifics' single use systems. *BioProcess Int. Industry Yearbook*: 134.

Matthews, T. et al. 2009. An Integrated Approach to Buffer Dilution and Storage. Pharmaceutical Manufacturing. http://www.pharmamanufacturing.com/articles/2009/046.html.

Mauter, M. 2009. Environmental life-cycle assessment of disposable bioreactors. *BioProcess Int.* 4: 18–29.

Mauter, M. 2009. Environmental life-cycle assessment of disposable bioreactors. *BioProcess Int.* 7(Suppl 4) :18–28.

Mauter, M. 2009. Environmental life-cycle comparison of conventional and disposable reactors. *GE Healthcare*.

Mazarevica, G., Diewok, J., Baena, J.R., Rosenberg, E., and Lendl, B. 2004. On-line fermentation monitoring by mid-infrared spectroscopy. *Appl. Spectrosc.* 58 (7): 804–810.

McArdle, J. 2004. Report of the workshop on monoclonal antibodies. *ATM* 32 (Suppl. 1): 119–122.

McDonald, K.A., Hong, L.M., Trombly, D.M., Xie, Q., and Jackman, A.P. 2005. Production of human α-l-antitrypsin from transgenic rice cell culture in a membrane bioreactor. *Biotechnol. Prog.* 21: 728–734.

Meissner. 2009. SALTUS—The Disposable Mixing and Bio-Reactor System. Brochure.

Menter, F.R. 1993. Zonal Two Equation k-ω Turbulence Models for Aerodynamic Flows. AIAA Paper: 93–2906.

Menzel, C., Lerch, T., Scheper, T., and Schugerl, K. 1995. Development of biosensors based on an electrolyte isolator semiconductor (EIS)-capacitor structure and their application for process monitoring. Part I. Development of the biosensors and their characterization. *Anal. Chim. Acta* 317 (1–3): 259–264.

Meyeroltmanns, F., Schmitz, J., and Nazlee, M. 2005. Disposable bioprocess components and single-use concepts for optimized process economy in biopharmaceutical production. *BioProcess Int.* 3: 60–66.

Migita, S., Ozasa, K., Tanaka, T., and Haruyama, T. 2007. Enzyme-based field-effect transistor for adenosine triphosphate (ATP) sensing. *Anal. Sci.* 23 (1): 45–48.

Mikola, M., Seto, J., and Amanullah, A. 2007. Evaluation of a novel Wave Bioreactor cell bag for aerobic yeast cultivation. *Bioprocess Biosyst. Eng.* 30: 231–241.

Millipore Corp. 2005. Lit. No. DS1002 EN00. Available: http://www.millipore.com/catalogue/module/c9423. Millipore. 2008. Mobius MIX500 disposable mixing system characterization. Billerica, Brochure.

Millipore Corp. 2009. Datasheet Mobius® CellReady 3 L Bioreactor. Available at: http://www.millipore.com/publications.nsf/ a73664f9f981af8c852569b9005b4 eee/228eeedd2285 ebe1852575de00570375/$FILE/DS26770000.pdf.

Millipore Corp. 2009. Mobius Disposable Mixing Systems. Billerica, Brochure.

Millipore Corp. 2010. Lynx S2S—Product Description. Available: http://www.millipore.com/catalogue/module/c9502..

Millipore Corp. 2010. Lynx ST Connectors—Data Sheet. Available: http://www.millipore.com/publications.nsf/a73664f9f981af8c852569b9005b4eee/402ffe097a6b Idca85256d510043ba64/$FILE/DS1750EN00.pdf.

Millipore Corp. 2010. NovaSeptum Sampling System—Product description. Available: http://www.millipore.com/catalogue/module/c10713.

Mills, A., Chang, Q., and Mcmurray, N. 1992. Equilibrium studies on colorimetric plastic film sensors for carbon dioxide. *Anal. Chem.* 64 (13): 1383–1389.

Mitchell, D.L. 1988. The relative cytotoxicity of (6-4): Photoproducts and cyclobutane dimers in mammalian cells. *Photochem. Photobiol.* 48: 51–57.

Mitchell, J.P., Court, J., Mason, M.D., Tabi, Z., and Clayton, A. 2008. Increased exosome production from tumor cell cultures using the Integra CELLine culture system. *J. Immunol, Methods* 335: 98–105.

Mohri, S., Shimizu, J., Goda, N., Miyasaka, T., Fujita, A., Nakamura, M., and Kajiya, F. 2006. Measurements of CO_2, lactic acid and sodium bicarbonate secreted by cultured cells using a flow-through type pH/CO_2 sensor system based on ISFET. *Sens. Actuators B Chem.* 115 (1): 519–525.

Monge, M. 2002. Single-use bag manifolds: Applications. *World Pharmaceutical Developments*, 85–86. www.sartorius.or.kr/nh/fileadmin/sartorius_pdf/.../manifold.pdf.

Monge, M. 2006. Successful project management for implementing single-use bioprocessing systems. *BioPharm Int.*; http://www.biopharminternational.findpharma.com/ biopharm/Article/Successful-Project-Management-for-Implementing-Sin/ ArticleStandard/Article/detail/423544.

Monge, M. 2008. Disposables: Process Economics—Selection, Supply Chain, and Purchasing Strategies. BioProcess International Conference. Anaheim, CA.

Mora, J., Sinclair, A., Delmdahl, N., and Gottschalk, U. 2006. Disposable membrane chromatography: Performance analysis and economic cost model. *BioProcess Int.* 4 (Suppl. 4): 38–43.

Morrow, K. 2006. Disposable bioreactors gaining favor. *Genet. Eng. Biotechnol. News* 26 (12): 42–45.

Mukherjee, J., Lindemann, C., and Scheper, T. 1999. Fluorescence monitoring during cultivation of *Enterobacter aerogenes* at different oxygen levels. *Appl. Microbiol. Biotechnol.* 52 (4): 489–494.

Muller, C., Hitzmann, B., Schubert, F., and Scheper, T. 1997. Optical chemo- and biosensors for use in clinical applications. *Sens. Actuators B Chem.* 40 (1): 71–77.

Muller, N., Girard, P., Hacker, D., Jordan, M., and Wurm, F.M. 2004. Orbital shaker technology for the cultivation of mammalian cells in suspension. *Biotechnol. Bioeng.* 89: 400–406.

Muller, W., Anders, K.D., and Scheper, T. 1989. Culture fluorescence measurements on immobilized yeast. *Chemie Ingenieur Technik* 61 (7): 564–565.

Muller, W., Wehnert, G., and Scheper, T. 1988. Fluorescence monitoring of immobilized microorganisms in cultures. *Anal. Chim. Acta* 213 (1–2): 47–53.

Munkholm, C., Walt, D.R., and Milanovich, F.P. 1988. A fiberoptic sensor for CO2 measurement. *Talanta* 35 (2): 109–112.

Murphy, T. and Schmidt, J. Animal-derived agents in disposable systems. *Gen. Eng. News.* 25(14); http://www.genengnews.com/articles/chtitem. aspx?tid=1090&chid=3.

Nagel, A., Koch, S., Valley, U., Emmrich, F., and Marx, U. 1999. Membrane-based cell culture systems—an alternative to in vivo production of monoclonal antibodies. *Dev. Biol. Stand.* 101: 57–64.

Nauman, E.B. 2004. Residence time distributions. In E.L. Paul, V.A. Atiemo-Obeng, S.M. Kresta (eds.), *Handbook of Industrial Mixing-Science and Practice.* Hoboken, NJ: John Wiley & Sons, 1–18.

Negrete, A. and Kotin, R.M. 2007. Production of recombinant adeno-associated vectors using two bioreactor configurations at different scales. *J. Virol.* 145: 155–161.

Negrete, A. and Kotin, R.M. 2008. Large-scale production of recombinant adeno-associated viral vectors. In J.M. Le Doux (ed.), *Gene Therapy Protocols: Vol. I, Series: Methods in Molecular Biology*, Vol. 433. Totowa, NJ: Humana Press, 79–96.

Nelson, K., Bielicki, J., and Anson, D.S. 1997. Immobilization and characterization of a cell line exhibiting a severe multiple sulphatase deficiency phenotype. *Biochem. J.* 326: 125–130.

Nguyen, H.J. 2009. Development of a membrane adsorber direct capture step: Process comparison of two commercial membranes. Presentation given at Sartorius Stedim Biotech Downstream Technology Forum, Berkeley, CA, October I.

Nicklin, D.J., Wilkes, J.O., and Davidson, J.F. 1962. Two-phase flow in vertical tubes. *Trans. Inst. Chem. Engrs.* 40: 61–68.

Nienow, A.W. 2006. Reactor engineering in large scale animal cell culture. *Cytotechnolgy* 50: 9–33.

Norling, L., Lute, S., Emery, R., Khuu, W., Voisard, M., Xu, Y., Chen, Q., Blan, G., and Brorson. K. 2005. Impact of multiple reuse of anion exchange chromatography media on virus removal. *J. Chromatogr.* I069: 79–89.

Norris, B.J., Gramer, M.J., and Hirschek, M.D. 2000. Growth of cell lines in bioreactors. In G.C. Howard, D.R. Bethell (eds.), *Basic Methods in Antibody Production and Characterization.* Boca Raton, FL: CRC Press, 87–104.

Norwood, D., Bail, D., Blanchard, J., Celado, L., Deng, T.J., De Grazio, F., Doub, B., Feinberg, T., Hendricker, A., Hrkach, J., McClellan, R., McGovern, T., Paskiet, D., Porter, D., Ruberto, M., Schroeder, A., Vogel, M., Wang, Q., Wolff, R., Munos, M., and Nagao, L. 2006. PQRI Leachables and Extractables Working Group. Safety thresholds and best practices for extractables and leachables in orally inhaled and nasal drug products. Product Quality Research Institute, Arlington VA, to the U.S. FDA Leachables and Extractables Working Group.

Norwood, D., Paskiet, D., Ruberto, M., Feinberg, T., Schroeder, A., Poochikian, G., Wang, Q., Deng, T.J., DeGrazio, F., Munos, M., and Nagao, L. 2007. Best practices for extractables and leachables in orally inhaled and nasal drug products: An overview of the PQRI recommendations. *Pharm. Res.* 4: 727–739.

Norwood, D.L. et al. 1995. Analysis of polycyclic aromatic hydrocarbons in metered dose inhaler drug formulations by isotope dilution gas chromatography/mass spectrometry. *J. Pharm. Biomed. Anal.* 13(293): 293–304.

Nova Biomedical Corporation. 2010. BioProfile, Automated Chemistry Analyzer. Available: http://www.novabiomedical.com.

Nova Biomedical Corporation. 2010. BioProfile, Automated Chemistry Analyzer. Available: http://www.novabiomedical.com. Albis Technologies AG. 2010. Zurich, Switzerland.

Novais, J.L., Titchener-Hooker, N.J., and Hoare, M. 2001. Economic comparison between conventional and disposables-based technology for the production of biopharmaceuticals. *Biotechnol. Bioeng.* 75 (2): 143–153.

Oashi, R., Singh, V., and Hamel, J.F.P. 2001. Perfusion culture in disposable bioreactors. *GEN* 21(40): 78.

Odian, G. 1981. *Principles of Polymerization*, Second Edition. New York: Wiley-Interscience Publication.

Oelßner, W., Zosel, J., Guth, U., Pechstein, T., Babel, W., Connery, J.G., Demuth, C., Grote Gansey, M., Verburg, J.B. 2005. Encapsulation of ISFET sensor chips. *Sens. Actuators B Chem.* 105 (1): 104–117.

Okonkowski, J., Balasubramanian, U., Seamans, C., Fischrogen, Z., Zhang, J., Lachs, P., Robinson, D., Chartrain, M. 2007. Cholesterol delivery to NS0 cells: challenges and solutions in disposable linear low-density polyethylene-based bioreactors. *J. Biosci. Bioeng.* 103: 50–59.

Öncül, A.A., Kalmbach, A., Genzel, Y., Reichl, U., and Thèvenin, D. 2009. Numerische und experimentelle Untersuchung der Fliessbedingungen in Wave-Bioreaktoren. *CIT* 81: 1241.

Outlook. 2008. Tufts Center for the Study of Drug Development, Boston.

Ozmotech Pty Ltd (Australia). Ozmotech Energy Technology. www.ozmoenergy.com/technology.

Ozturk, S.S. 2007. Comparison of product quality: Disposable and stainless steel bioreactor. BioProduction 2007, Berlin, Germany.

Pahl, M. 2003. Mischtechnik, Aufgaben und Bedeutung. In Kraume, M. (ed.), *Mischen und Rühren-Grundlagen und moderne Verfahren*. Weinheim: Wiley-VCH Verlag GmbH & Co. KGaA, 1–19.

Palazon, J., Mallol, A., Eibl, R., Lettenbauer, C., Cusidó, R.M., and Piñol, M.T. 2003. Growth and ginsenoside production in hairy root cultures of *Panax ginseng* using a novel bioreactor. *Planta Med.* 69: 344–349.

Palazon, J., Mallol, A., Eibl, R., Lettenbauer, C., Cusido, R.M., and Pinyol, M.T. 2003. Growth and ginsenoside production in hairy root cultures of *Panax ginseng* using a novel bioreactor. *Planta Med* 69: 344–349.

Pall. 2010. Kleenpak Sterile Connectors—Product description. Available: http://www.pall.com/variants/pdf/pdf/biopharm_34125.pdf.

Pandurangappata, M., Lawrence, N.S., Jiang, L., Jones, T.G., and Compton, R.G. 2003. Physical adsorption of N,N'-diphenyl-p-phenylenediamine onto carbon particles: Application to the detection of sulfide. *Analyst* 128 (5): 473–479.

Pang, J., Blanc, T., Brown, J., Labrenz, S., Villalobos, A., Depaolis, A., Gunturi, S., Grossman, S., Lisi, P., and Heavner, G.A. 2007. Recognition and identification of UV absorbing leachables in EPREX® pre-filled syringes: An unexpected occurrence at a formulation-component interference. *PDA J. Pharm. Sci. Technol.* 61: 423–432.

Parenteral Drug Association. 2007. Technical report no. 1. *J. Pharm. Sci. Technol.* 61 (Suppl.): S-1.

Parkinson, S. 2009. User's Guide Thermo Scientific Hyclone single-use mixer. User guide, UG004 Rev4.

Paschedag, A.R. 2004. *CFD in der Verfahrenstechnik*. Weinheim: Wiley-VCH.

Patterson, B.J. 2009. A Closer Look at Automated In-Line Dilution. Available: http://pharmtech.findpharma.com/pharmtech/article/articleDetail.jsp?id=632935&sk=&date=&pageID=4.

Paul, E.L., Atiemo-Obeng, V.A., and Kresta, S.M. 2004. Introduction. In Paul, E.L., Atiemo-Obeng, V.A., and Kresta, S.M. (eds.), *Handbook of Industrial Mixing-Science and Practice*. Hoboken, NJ: John Wiley & Sons, xxxiii–lviii.

PDA Technical Report No. 26. 1998. Sterilizing Filtration of Liquids. *PDA J. Pharmaceut. Sci. Technol.* 52, S1.

PDC Aseptic Filling Systems. 2010. Thermoelectric Tube Sealer—Technical data sheet. Available: http://www.asepticfilling.com/Thermoelectric%20Tube%20 Sealer%20Data%20Sheet%20Reviewed%20Rev%205.pdf.

Peacock, L. and Auton, K.A. 2008. Comparing shaker flasks with a single-use bioreactor for growing yeast seed cultures. *BioProcess Int.* 6: 54–57.

Pekeler, T., Lindemann, C., Scheper, T., and Hitzmann, B. 1998. Prediction of bioprocess parameters from two-dimensional fluorescence spectra. *Chemie Ingenieur Technik* 70 (12): 1610–1611.

PendoTECH. 2009. Process scale single use pressure sensors. Available: http://www. pendotech.com/products/disposable_pressure_sensors/disposable_pressure_ sensors.htm.

Peter, C.P., Suzuki, Y., and Büchs, J. 2006. Hydromechanical stress in shake flask: Correlation for the maximum local energy dissipation rate. *Biotechnol Bioeng* 93: 1164–1176.

Pharmaceutical Inspection Co-operation Scheme. 2007. Aide memoire: Inspection of biotechnology manufacturers. PI 024–2, PIC/S Secretariat. Geneva, Switzerland.

Pharmaceutical Inspection Co-operation Scheme. 2007. Isolators used for aseptic processing and sterility testing. PI 014–3, PIC/S Secretariat. Geneva, Switzerland.

Pharmaceutical Inspection Co-operation Scheme. 2007. Recommendations on validation master plan, installation and operational, nonsterile process validation, cleaning validation. PI 006–3, PIC/S Secretariat. Geneva, Switzerland.

Pharmaceutical Inspection Co-operation Scheme. 2007. Recommendation on sterility testing. PI 012–3, PIC/S Secretariat Geneva, Switzerland.

Pharmaceutical Inspection Co-operation Scheme. 2009. Recommendations on the validation of aseptic processes. PI 007–5, PIC/S Secretariat. Geneva, Switzerland.

Phillips, M., Cormier, J., Ferrence, J., Dowd, C., Kiss, R., Lutz, H., and Carter, J. 2005. Performance of a membrane adsorber for trace impurity removal in biotechnology manufacturing. *Chromatogr. A* 1078: 74–82.

Pierce, L.N. and Shabram, P.W. 2004. Scalability of a disposable bioreactor from 25 L–500 L run in perfusion mode with a CHO cell-based cell line: A tech review. *BioProcessing J* 4: 51–56.

Pinto, F. 1999. Effect of experimental parameters on plastics pyrolysis reactions. Proceedings R99. Geneva.

Pora, H. 2006. Increasing bioprocessing efficiency, single use technologies. *Pharm. Technol. Eur.* 18: 24–29.

Pora, H. 2006. The case for disposable manufacturing equipment to accelerate vaccine development. *BioPharm Int.* 19(6).

Pora, H. and Rawlings, B. 2009. A user's checklist for introducing single-use components into process systems. *BioProcess Int.* 4: 9–16.

Pora, H. and Rawlings, B. 2009. Managing solid waste from single-use systems in biopharmaceutical manufacturing. *BioProcess Int.* 7(1): 18–25.

Potera, C. 2009. Firm on quest to improve biomanufacturing. *Genet Eng. Biotechnol. News* 7: 20–21.

PQRI Leachables and Extractables Working Group. Safety Thresholds and Best Practices for Extractables and Leachables in Orally Inhaled and Nasal Drug Products. Product Quality Research Institute: Arlington, VA; http://www.pqri.org/pdfs/LE_Recommendations_to_FDA_09-29-06.pdf. 2006.

Prashad, M. and Tarrach, K. 2006. Depth filtration aspects for the clarification of CHO cell derived biopharmaceutical feed streams. *FISE* 9: 28–30.

Press/BioPlan Associates, 1–28.

Product Quality Research Institute

Proulx, S.P. and Furey, J.F. 2007. Disposable, pre-sterilized fluid receptacle sampling device. 7293477 B2, Millipore Corporation, USA.

Purefit data sheet FLS-3309A-1.5M-1008–SGCS. Saint-Gobain Performance Plastics 2008.

Quattroflow. 2010. Catalogue. Available: http://www.quattroflow.com/055c079b 5a0c11315/index.html.

Rafa, B. and Panofen, F. 2009. Reprásentativ und kontaminations-frei. *P&A Biotech* 1: 37.

Ransohoff, T. 2004. Disposable chromatography: Current capabilities and future possibilities. BPD North Carolina Biotechnology Center, November 18.

Rao, G., Moreira, A., and Brorson, K. 2009. Disposable bioprocessing: The future has arrived. *Biotechnol. Bioeng.* 102(2): 348–356.

Rathore, N. and Rajan, R.S. 2008. Current perspectives on stability of protein drug products during formulation, fill and finish operations. *Biotechnol. Prog.* 24 (3): 504–514.

Rauth, A.M. 1965. The physical state of viral nucleic acid and the sensitivity of viruses to ultraviolet light. *Biophys. J.* 5: 257–273.

Raval, K., Liu, C.M., and Büchs, J. 2006. Large-scale disposable shaking bioreactors. *BioProcess Int.* 1: 46–49.

Ravise, A. et al. 2010. Hybrid and disposable facilities for manufacturing of biopharmaceuticals: Pros and cons. *Adv. Biochem. Eng. Biotechnol.* 115: 185–219. DOI: ID 1007/10_2008_24.

Rawlings, B. and Pora, H. 2009. Environmental impact of single-use and reusable bioprocess systems. *BioProcess Int.* 2: 18–25.

Ray, S. and Tarrach, K. 2008. Virus clearance strategy using a three-tier orthogonal technology platform. *BioPharm Int.* 22: 50–58.

Reactor development for the hepatitis A vaccine VAQTA. In: Carrondo, M.J.T., Griffiths, B., and Moreira, J.L.P. (eds.), *Animal Cell Technology: From Vaccine to Genetic Medicine.* Kluwer, Dordrecht, pp. 175–183.

Reardon, K.F., Scheper, T., Anders, K.D., Muller, W., and Buckmann, A.F. 1989. Novel applications of fluorescence sensors. 11th Symposium on Biotechnology for Fuels and Chemicals, Colorado Springs, CO.

Reardon, K.F., Scheper, T., and Bailey, J.E. 1986. In situ fluorescence monitoring of immobilized *Clostridium acetobutylicum. Biotechnol. Lett.* 8 (11): 817–822.

Reardon, K.F., Scheper, T.H., and Bailey, J.E. 1987. Use of a fluorescence sensor for measurement of NAD(P)H-dependent culture fluorescence of immobilized cell systems. *Chemie Ingenieur Technik* 59 (7): 600–601.

Reif, O.W., Solkner, P., and Rupp, J. 1996. Analysis and evaluation of filter cartridge extractables for validation of pharmaceutical downstream processing. *J. Pharm. Sci. Technol.* 50: 399–407.

Resin Identification Codes. American Chemistry Council; http://www.american chemistry.com/s_plastics/bin.asp?CID=1102&DID=4645&DOC=FILE.

Rhee, J.I., Lee, K.I., Kim, C.K., Yim, Y.S., Chung, S.W., Wei, J.Q., and Bellgardt, K.H. 2005. Classification of two-dimensional fluorescence spectra using self-organizing maps. *Biochem. Eng. J.* 22 (2): 135–144.

Ries, C., John, C., and Eibl, R. 2009. Einwegbioreaktoren für die Prozessentwicklung mit Insektenzellen. *Bioforum* 3: 11–13.

Ries, C. 2008. The process engineering characteristics of the Thermo Fisher Scientific Single-Use Bioreactor 50 L: Determination of mixing time, power input and kLa values. Application note. Available from Thermo Fisher Scientific.

Rios, M. 2006. Process considerations for cell-based influenza vaccines. *Pharm. Technol.* 4: 1–6.

Ritala, A., Wahlström, E.H., Holkeri, H., Hafren, A., Mäkelainen, K., Baez, J., Mäkinen, K., and Nuutila, A.M. 2008. Production of a recombinant industrial protein using barley cell cultures. *Protein Expr. Purif.* 59: 274–281.

Röll, M. 2006. Thermal welding for sterile connections. *Genet. Eng. Biotechnol. News* 26: 64.

Rombach, C. 2005. Single-use benefits and hopes. *BioProcess Int.* 3 (2): 88.

Roy, J. 2002. Pharmaceutical impurities—A mini-review. *AAPS Pharm. Sci. Technol.* 3: 1–6.

Royce, J. et al. 2008. Guidelines for Selecting Normal Flow Filters. Pharmaceutical Technology. Also available online at http://www.nxtbook.com/nxtbooks/advanstar.

Roychoudhury, P., Harvey, L.M., and McNeil, B. 2006. At-line monitoring of ammonium, glucose, methyl oleate and biomass in a complex antibiotic fermentation process using attenuated total reflectance-mid-infrared (ATR-MIR) spectroscopy. *Anal. Chim. Acta* 561: 218–224.

Rudolph, G., Bruckerhoff, T., Bluma, A., Korb, G., and Scheper T. 2007. Optical inline measurement procedure for cell count and cell size determination in bioprocess technology. *Chemie Ingenieur Technik* 79 (1–2): 42–51.

Ryder, M., Fisher, S., Hamilton, G., Hamilton, M., and James, G. 2007. Bacterial Transfer through Needlefree Connectors: Comparison of Nine Different Devices. Poster. Available: http://www.icumed.com/Docs-Clave/Ryder%20 SHEA%202007%20Poster.pdf.

Safety Thresholds and Best Practices for Extractables and Leachables in Orally Inhaled and Nasal Drug Products. Product Quality Research Institute: Arlington, VA. 2006.

Sagi, E., Hever, N., Rosen, R., Bartolome, A.J., Premkumar, J.R., Ulber, R., Lev, O., Scheper, T., and Belkin, S. 2003. Fluorescence and bioluminescence reporter functions in genetically modified bacterial sensor strains. *Sens. Actuators B Chem.* 90 (1–3): 2–8.

Sandstrom, C. and Schmidt, B. 2005. Facility-design considerations for the use of disposable bags. *BioProcess Int.* (Suppl. 4): 56–60.

Sartorius Stedim Biotech. 2000. Sacova Valve—Technical Information. Available: http://sartorius.or.kr/B_Braun_Biotech/Fermenters_and_Bioreactors/pdf/ TI_SACOVAe_02-00.pdf.

Sartorius Stedim Biotech. 2009. Flexel 3D LevMix system for Palletank. Göttingen, Brochure.

Sartorius Stedim Biotech. 2009. Flexel 3D system for recirculation mixing. Göttingen, Brochure.

Sartorius Stedim Biotech. 2009. White Paper: Evolving Toward Single-Use Bioprocessing: from Solutions to Holistic Value Creation. Bioresearch Online; http://www.bioresearchonline.com/download.mvc/Evolving-Toward -Single-Use-Bioprocessing-From-0001.

Sartorius Stedim Biotech. 2010. Aseptic Transfer System—Definition of the technology. Available: http://www.sartorius-stedim.com/WW/en/Aseptic-Transfer-Systems/Biosafe%C2%AE-Aseptic-Trans-Single-use-Bag/5eiax0td5dd/ brbua8nsl0k/mp.htm.

Sartorius Stedim Biotech. 2010. Opta SFT-I—Product description. Available: http:// www.sartorius.com/fileadmin/sartorius_pdf/alle/biotech/Data_Opta_SFT-l_ SLO2000-e.pdf.

Sato, K., Yoshida, Y., Hirahara, T., and Ohba, T. 2000. On-line measurement of intra-cellular ATP of Saccharomyces cerevisiae and pyruvate during sake mashing. *J. Biosci. Bioeng.* 90 (3): 294–301.

Schears, G., Schultz, S.E., Creed, J., Greeley, W.J., Wilson, D.F., and Pastuszko, A. 2003. Effect of perfusion flow rate on tissue oxygenation in newborn piglets during cardiopulmonary bypass. *Ann. Thorac. Surg.* 75 (2): 560–565.

Scheper, T., Brandes, W., Grau, C., Hundeck, H.G., Reinhardt, B., Ruther, F., Plotz, F., Schelp, C., Schugerl, K., Schneider, K.H., Giffhorn, F., Rehr, B., and Sahm, H. 1990. Applications of biosensor systems for bioprocess monitoring. 3rd International Symposium on Analytical Methods in Biotechnology (Anabiotec 90), San Francisco, CA.

Scheper, T., Brandes, W., Maschke, H., Ploetz, F., and Mueller C. 1993. Two FIA-based biosensor systems studied for bio-process monitoring. *J. Biotechnol.* 31 (3): 345–356.

Scheper, T. and Buckmann, A.F. 1990. A fiber optic biosensor based on fluorometric detection using confined macromolecular nicotinamide adenine-dinucleotide derivatives. *Biosens. Bioelectron.* 5 (2): 125–135.

Scheper, T., Gebauer, A., Kuhlmann, W., Meyer, H.D., and Schugerl, K. 1984. Dechema-Monographien 95.

Scheper, T., Gebauer, A., and Schugerl, K. 1987. Monitoring of NADH-dependent culture fluorescence during the cultivation *Escherichia coli*. *Chem. Eng. J.* 34 (1): B7–B12.

Scheper, T., Hitzmann, B., Stark, E., Ulber, R., Faurie, R., Sosnitza, P., and Reardon, K.F. 1999. Bioanalytics: Detailed insight into bioprocesses. *Anal. Chim. Acta* 400: 121–134.

Scheper, T., Lorenz, T., Schmidt, W., and Schugerl, K. 1987. Online measurement of culture fluorescence for process monitoring and control of biotechnological processes. *Ann. N.Y. Acad. Sci.* 506: 431–445.

Scheper, T. and Schugerl, K. 1986. Bioreactor characterization by in situ fluorometry. *Chemie Ingenieur Technik* 58 (5): 433–433.

Scheper, T. and Schugerl, K. 1986 (February). Characterization of bioreactors by in situ fluorometry. *J. Biotechnol.* 3 (4): 221–229.

Scheper, T. and Schugerl, K. 1986. Culture fluorescence studies on aerobic continuous cultures of *Saccharomyces cerevisiae*. *Appl. Microbiol. Biotechnol.* 23 (6): 440–444.

Scheper, T. 1992. Biosensors for process monitoring. *J. Ind. Microbiol.* 9 (3–4): 163–172.

Scheper, T.H., Hilmer, J.M., Lammers, F., Muller, C., and Reinecke M. 1996. Biosensors in bioprocess monitoring. *J. Chromatogr. A* 725 (1): 3–12.

Scheper, T. and Jornitz, M.W. (eds.). 2006. Sterile filtration. *Advances in Biochemical Engineering/Biotechnology*. Berlin: Springer.

Schmidt, S., and Kauling, J. 2007. Process and laboratory scale UV inactivation of viruses and bacteria using to innovative coiled tube reactor. *Chem. Eng. Technol.* 30: 945–950.

Schmidt, S., Mora, J., Dolan, S., and Kauling, J. 2005. An integrated concept for robustly and efficient virus clearance and contaminant removal in biotech processes. *BioProcess Int.* 8: 26–31.

Schneditz, D., Kenner, T., Heimel, H., and Stabinger, H. 1989. A sound-speed sensor for the measurement of total protein-concentration in disposable, blood-perfused tubes. *J. Acoust Soc. Am.* 86 (6): 2073–2080.

Schöning, M.J., Brinkmann, D., Rolka, D., Demuth, C., and Poghossian, A. 2005. CIP (cleaning-in-place) suitable non-glass pH sensor based on a Ta2O5-gate EIS structure. *Sens. Actuators B Chem.* 111–112: 423–429.

Schreyer, H.B., Miller, S.E., and Rodgers, S. 2007. Application note: High-throughput process development. *Genet Eng. Biotechnol. News*. Available: http://www.genengnews.com/issues.com/issues/item.aspx?issue_id=78.

Schugerl, K., Bellgardt, K.H., Kretzmer, G., Hitzmann, B., and Scheper, T. 1993. In-situ and online monitoring and control of biotechnological processes. *Chemie Ingenieur Technik* 65 (12): 1447–1456.

Schugerl, K., Lindemann, C., Marose, S., and Scheper, T. 1998. Bioprocess Engineering Course, p. 400.

Schugerl, K., Lorenz, T., Lubbert, A., Niehoff, J., Scheper, T., and Schmidt, W. 1986. Pros and cons—On-line versus off-line analysis of fermentations. *Trends Biotechnol.* 4 (1): 11–15.

Schugerl, K., Lubbert, A., and Scheper, T. 1987. Online measurement of bioreactor performance. *Chemie Ingenieur Technik* 59 (9): 701–714.

Schugerl, K. 2001. Progress in monitoring, modeling and control of bioprocesses during the last 20 years. *J. Biotechnol.* 85 (2): 149–173.

Schwan, S., Fritzsche, M., Cismak, A, Heilmann, A., and Spohn, U. 2007. In vitro investigation of the geometric contraction behavior of chemo-mechanical P-protein aggregates (forisomes). *Biophys. Chem.* 125 (2–3): 444–452.

Schwander, E. and Rasmusen, H. 2005. Scalable, controlled growth of adherent cells in a disposable, multilayer format. *Genet. Eng. Biotechnol. News* 25: 29.

Scientific HyClone BPCs® Products and Capabilities. 2008/2009. Catalog.

SciLog. 2009. Disposable, pre-calibrated SciCon conductivity sensors. Available: www.scilog.com/sensor/conductivity.php.

SciLog. 2009. Pressure Sensors & Monitors. Available: http://www.scilog.com/sensor/pressure.php.

Scott, C. 2007. "Single-use" doesn't necessarily mean "disposable." *BioProcess Int.* 5(Suppl. 4): 4.

Scott, C. 2007. Disposables qualification and process validation. *BioProcess Int.* 5: 24–27.

Scott, C. 2007. Single-use bioreactors: A brief review of current technology. *BioProcess Int.* 5 (Suppl. 5): 44–51.

Scott, C. 2008. Biotech leads a revolution in vaccine manufacturing. *BioProcess Int.* 6 (Suppl. 6): 12–18.

Scott, L.E., Aggett, H., and Glencross, D.K. 2001. Manufacture of pure antibodies by heterogeneous culture without downstream purification. *Biotechniques* 31: 666–668.

SEBRA. 2010. Aseptic Sterile Welder—Product description. Available: http://www.sebra.com/BCP-3960.html.

SEBRA. 2010. Tube Sealer—Product descriptions. Available: http://www.sebra.com/BCP-biopharmaceutical.html.

Selvanayagam, Z.E., Neuzil, P., Gopalakrishnakone, P., Sridhar, U., Singh, M., and Ho, L.C. 2002. An ISFET-based immunosensor for the detection of [beta]-Bungarotoxin. *Biosens. Bioelectron.* 17 (9): 821–826.

Severinghaus, J.W. and Bradley, A.F. 1958. Electrodes for blood pO_2 and pCO_2 determination. *J. Appl. Physiol.* 13 (3): 515–520.

Sevilla, F., Kullick, T., and Scheper, T. 1994. A bio-FET sensor for lactose based on co-immobilized α-galactosidase/glucose dehydrogenase. *Biosens. Bioelectron.* 9 (4–5): 275–281.

Sharma, B., Bader, F., Templeman, T., Lisi, B., Ryan, M., and Heavner, G.A. 2004. Technical investigations into the cause of the increased incidence of antibody-mediated pure red cell aplasia associated with EPREX®. *Eur. J. Hosp. Pharm.* 5: 86–91.

Shire, S.J. 2009. Formulation and manufacturability of biologics. *Curr. Opin. Biotechnol.* 6: 708–714.

Shukla, A., Hubbard, B., Tressel, T., Guhan, S., and Low, D. 2007. Downstream processing of monoclonal antibodies—Application of platform approaches. *J. Chromatogr. B* 828: 28–39.

Simonet, J. and Gantzer, C. 2006. Inactivation of poliovirus I and F-specific RNA phages and degradation of their genomes by UV irradiation at 254 nanometers. *Appl. Environ. Microbiol.* 72: 7671–7677.

Simonis, A., Dawgul, M., Luth, H., and Schöning, M.J. 2005. Miniaturised reference electrodes for field-effect sensors compatible to silicon chip technology. *Electrochim. Acta* 51 (5): 930–937.

Simonis, A., Luth, H., Wang, J., and Schöning, M.J. 2004. New concepts of miniaturised reference electrodes in silicon technology for potentiometric sensor systems. *Sens. Actuators B Chem.* 103 (1–2): 429–435.

Sinclair, A. et al. 2008. The environmental impact of disposables technologies. *BioPharm Int.*, supplement; http://www.biopharminternational.findpharma.com/biopharm/article/articleDetail.jsp?id=566014&pageID=1&sk=&date=.

Sinclair, A., Leveen, L., Monge, M., Lim, J., and Cox, S. 2008. The environmental impact of disposable technologies. *BioPharm. Int.* 11: 1–11.

Sinclair, A. and Monge, M. 2002. Quantitative economic evaluation of single use disposables in bioprocessing. *Pharm. Eng.* 22: 20–34.

Sinclair, A. and Monge, M. 2004. Biomanufacturing for the 21st century: Designing a concept facility based on single-use systems. *BioProcess Int.* 2: 26–31.

Sinclair, A. and Monge, M. 2005. Concept facility based on single antibody manufacture. *J. Chromatogr. B* 8–18: 848.

Sinclair, A. and Monge, M. 2005. Concept facility based on single-use systems: Part 2. *BioProcess Int.* 3(9): S51–S55.

Sinclair, A. and Monge, M. 2009. Disposables cost contributions: A sensitivity analysis. *BioPharm Int.* 22(4): 14–18.

Sinclair, A. and Monge, M. 2009. Evaluating disposable mixing systems. *Biopharm Int.* 22 (2): 24–29.

Sinclair, A. 2009. Biological products manufacturing: Cost challenges and opportunities now and in the future. BPI, Raleigh, NC, October 12.

Sinclair, A., Leveen et al. 2008. The Environmental Impact of Disposable Technologies, The Biopharm International Guide, November 2008; Base of the analysis: Typical mAb process at 3 x 2000 L scale.

Sinclair, A. and Monge, M. 2004. Biomanufacturing for the 21st century: Designing a concept facility based on single-use systems. *BioProcess Int.* 2 (Suppl.): 26–31.

Singh, S.K., Kolhe, P., Wang, W., and Nema, S. 2009. Large-scale freezing of biologics: A practitioner's review, part 1: Fundamental aspects. *BioProcess Int.* 7 (9): 32–44.

Singh, S.K., Kolhe, P., Wang, W., and Nema, S. 2009. Urge-scale freezing of biologics: A practitioner's review, part 2: Practical advice. *BioProcess Int.* 7 (10): 34–42.

Singh, V. 1999. Disposable bioreactor for cell culture using wave-induced motion. *Cytotechnology* 30: 149–158.

Singh, V. 2004. Bioprocessing tutorial: Mixing in large disposable containers. *Genet. Eng. Biotechnol. News* 24 (3): 42–43.

Singh, V. 2005. *The Wave Bioreactor Story.* Wave Biotech LLC, Somerset.

Slivac, I., Srček, V.G., Radoševic, K., Kmetič, I., and Kniewald, Z. 2006. Aujeszky's disease virus production in disposable bioreactors. *J. Biosci.* 3: 363–368.

Smiley, D., Rhee, M., Ziemer, D., Gallina, D., Phillips, L.S., Kolm, P., and Umpierrez, G.E. 2005. Lack of follow-up care: A major obstacle to optimal care in Latinos with diabetes. *Diabetes* 54: A588–A589.

Smith, M. 2004. An evaluation of Protein A and non-Protein A methods for the recovery of monoclonal antibodies and considerations for process scale-up. Scaling-up of Biopharmaceutical Products, The Grand, Amsterdam, January.

Smith, M. 2005. Strategies for the purification of high titer, high volume mammalian cell culture batches. BioProcess International European Conference and Exhibition, Berlin, April.

Sofer, G. and Lister, D.C. 2003. Inactivation methods grouped by virus: Virus inactivation in the 1990s and into the 21st century. *BioPharm. Int* 6: 37–42.

Solle, D., Geissler, D., Stark, E., Scheper, T., and Hitzmann, B. 2003. Chemometric modelling based on 2D-fluorescence spectra without a calibration measurement. *Bioinformatics* 19 (2): 173–177.

Song, K.-S., Zhang, G.-J., Nakamura, Y., Furukawa, K., Hiraki, T., Yang, J.-H., Funatsu, T., Ohdomari, I., and Kawarada, H. 2006. Label-free DNA sensors using ultrasensitive diamond field-effect transistors in solution. *Phys. Rev. E Stat. Nonlin. Soft Matter Phys.* 74 (4): 041919.

Stärk, E., Hitzmann, B., Schügerl, K., Scheper, T., Fuchs, C., Köster, D., and Märkl, H. 2002. In-situ-fluorescence-probes: A useful tool for non-invasive bioprocess monitoring. *Adv. Biochem. Eng. Biotechnol.* 74: 21–38.

Stein, A. and Kiesewetter, A. 2007. Cation exchange chromatography in antibody purification: pH screening for optimized binding and HCP removal. *J. Chromatogr. B* 848: 151–158.

Stone, T.E., Goel, V., and Leszczak, J. 1994. Methodology for analysis of extractables: A model stream approach. *Pharm. Technol.* 18: 116–121.

Streeter, I., Leventis, H.C., Wildgoose, G.G., Pandurangappa, M., Lawrence, N.S., Jiang, L., Jones, T.G.J., and Compton, R.G. 2004. A sensitive reagentless pH probe with a ca. 120 mV/pH unit response. *J. Solid State Electrochem.* 8 (10): 718–721.

Suhr, H., Wehnert, G., Schneider, K., Bittner, C., Scholz, T., Geissler, P., Jahne, B., and Scheper, T. 1995. In-situ microscopy for online characterization of cell-populations in bioreactors, including cell-concentration measurements by depth from focus. *Biotechnol. Bioeng.* 47 (1): 106–116.

Sukumar, M. and Brorson, K. 2007. Phage passage after extensive processing in small-virus retentive filters. *Biotechnol. Appl. Biochem.* 47: 141–152.

Surribas, A., Geissler, D., Gierse, A., Scheper, T., Hitzmann, B., Montesinos, J.L., and Valero, F. 2006. State variables monitoring by in situ multi-wavelength fluorescence spectroscopy in heterologous protein production by *Pichia pastoris*. *J. Biotechnol.* 124 (2): 412–419.

Szalai, E.S., Alvarez, M.M., and Muzzio, F.J. 2004. Laminar mixing: A dynamic systems approach. In E.L. Paul, V.A. Atiemo-Obeng, S.M. Kresta (eds.), *Handbook of Industrial Mixing-Science and Practice*. Hoboken, NJ: John Wiley & Sons, 89–144.

Tallentire, A. 1980. The spectrum of microbial radiation sensitivity. *Radiat. Phys. Chem.* 15: 83–89.

Tamachi, T., Maezawa, Y., Ikeda, K., Kagami, S., Hatano, M., Seto, Y., Suto, A., Suzuki, K., Watanabe, N., Saito, Y., Tokihisa, T., Iwamoto, I., and Nakajima, H. 2006. IL-25 enhances allergic airway inflammation by amplifying a TH2-cell dependent pathway in mice. *J. Allergy Clin. Immunol.* 118: 606–614.

Tan, W.H., Shi, Z.Y., and Kopelman, R. 1992. Development of submicron chemical fiber optic sensors. *Anal. Chem.* 64 (23): 2985–2990.

Tarrach, K., Meyer, A., Dathe, J.E., and Sunn, H. 2007. The effect of flux decay on a 20 nm nanofilter for virus retention. *BioPharm. Int* 4: 58–63.

Tarrach, K. 2005. Integrative strategies for viral clearance. 4th Annual Biological Production Forum, Edinburgh, UK.

Tarrach, K. 2007. Process economy of disposable chromatography in antibody manufacturing: Development and production of antibodies, vaccines, and gene vectors. WilBio's Bioprocess Technology, Amsterdam, The Netherlands.

Tarrach, K. 2007. Virus filter positioning in the purification process of cell culture intermediates and flow decay aspects associated with small non-enveloped virus retention. Bioprocess Internationally European Conference and Exhibition, Paris, France.

Tartakovsky, B., Sheintuch, M., Hilmer, J.M., and Scheper, T. 1996. Application of scanning fluorometry for monitoring of a fermentation process. *Biotechnol. Prog.* 12 (1): 126–131.

Taylor, I. 2007. The CellMaker plus single-use bioreactor: A new bioreactor capable of culturing bacteria, yeast, insect and mammalian cells. Hannover: Biotechnica, Teixeira, A.P., Portugal, C.A.M., Carinhas, N., Dias, J.M.L., Crespo, J.P., Alves, P.M., Carrondo, M.J.T., and Oliveira, R. 2009. In situ 2D fluorometry and chemometric monitoring of mammalian cell cultures. *Biotechnol. Bioeng.* 102 (4): 1098–1106.

Terrier, B., Courtois, C., Hénault, N., Cuvier, A., Bastin, M., Aknin, A., Dubreuil, J., and Pétiard, V. 2007. Two new disposable bioreactors for plant cell cultures: The wave & undertow bioreactor and the slug bubble bioreactor. *Biotechnol. Bioeng.* 96: 914–923.

Terumo. 2010. Sterile Tubing Welders—Website. Available: http://www.terumo transfusion.com/ProductCategory.aspx?categoryId=6.

Terumo. 2010. Teruseal Tube Sealer—Product description. Available: http://www.terumotransfusion.com/ProductDetails.aspx?categoryld=5. The United States Pharmacopeia. 2008. USP 31, NF 26. Rockville, IN.

The Irradiation and Sterilization Subcommittee of the Bio-Process Systems Alliance. 2007. Guide to irradiation and sterilization validation of single-use bioprocess systems, Part 2. *BioProcess Int.* 5(10): 60–70.

Thermo Fisher Scientific. 2009. Thermo Scientific Nalgene Bioprocess Bag Management System. Available: http://www.nalgenelabware.com/features/featureDetail.asp?featureID=70.

Thoemmes, J. and Kula, M. 1995. Membrane chromatography: An integrative concept in the downstream processing of proteins. *Biotechnol. Prog. I* 1: 357–367.

Thommes, J. and Etzel, M. 2007. Alternatives to chromatographic separations. *Biotechnol. Prog.* 23: 42–45.

Thordsen, O., Lee, S.J., Degelau, A., Scheper, T., Loos, H., Rehr, B., and Sahm, H. 1993. A model system for a fluorometric biosensor using permeabilized *Zymomonas mobilis* or enzymes with protein confined dinucleotides. *Biotechnol. Bioeng.* 42 (3): 387–393.

Tollnik, C. 2009. Einsatz von Disposables in der Praxis—ein Erfahrungsbericht zu Design und Betrieb einer Pilotanlage für klinische Wirkstoffproduktionen. 2. Konferenz Einsatz von Single-Use-Disposables (Concept Heidelberg). Mannheim, Germany.

Trebak, M., Chong, J.M., Herlyn, D., and Speicher, D.W. 1999. Efficient laboratory-scale production of monoclonal antibodies using membrane-based high-density cell culture technology. *J. Immunol. Methods* 230: 59–70.

Trevisan, M.G. and Poppi, R.J. 2008. Direct determination of ephedrine intermediate in a biotransformation reaction using infrared spectroscopy and PLS. *Talanta* 75: 1021–1027.

Tservistas, M., Koneke, R., Comte, A., and Scheper, T. 2001. Oxygen monitoring in supercritical carbon dioxide using a fibre optic sensor. *Enzyme Microb. Technol.* 28 (7–8): 637–641.

Tutorial. 2005. High-yield single-use cell culture systems. *Genet Eng. Biotechnol. News.* Available: http://www.genengnews.com/articles/chtitem.aspx?tid=1093&chid=3.

Ulber, R., Frerichs, J.G., and Beutei, S. 2003. Optical sensor systems for bioprocess monitoring. *Anal. Bioanal. Chem.* 376 (3): 342–348.

Ulber, R., Hitzmann, B., and Scheper, T. 2001. Innovative bio-process analysis—New approaches to understanding biotechnological processes. *Chemie Ingenieur Technik* 73 (1–2): 19–26.

Ulber, R., Protsch, C., Solle, D., Hitzmann, B., Willke, B., Faurie, R., and Scheper, T. 2001. Use of bioanalytical systems for the improvement of industrial tryptophan production. *Chem. Eng. Technol.* 24 (7): 15–17.

Ulber, R. and Scheper, T. 1998. Enzyme biosensors based on ISFETs. In Enzyme and Microbial Biosensors. Mulchandani, A. and Rogers, K.R. (eds.). Berlin: Springer, 35–50.

U.S. Department of Health and Human Services. 1987. *Guidance for Industry: Guideline for Validation of Limulus amebocyte lysate Test as an End-Product Endotoxin Test for Human and Animal Parenteral Drugs, Biological Products and Medicinal Devices.* Rockville, IL: Food and Drug Administration.

U.S. Department of Health and Human Services. 1998. *Guidance for Industry: Manufacturing, Processing or Holding Active Pharmaceutical Ingredients.* Rockville, IL: Food and Drug Administration.

U.S. Department of Health and Human Services. 1999. *Guidance for Industry: Container Closure Systems for Packaging Human Drugs and Biologies—Chemistry, Manufacturing and Controls Documentation.* Rockville, IL: Food and Drug Administration.

U.S. Department of Health and Human Services. 2004. *Guidance for Industry: Sterile Drug Products Produced by Aseptic Processing—Current Good Manufacturing Practice.* Rockville, IL: Food and Drug Administration.

U.S. Department of Health and Human Services. 2004. *Guidance for Industry: PAT Process Analytical Technology—A Framework for Innovative Pharmaceutical Manufacturing and Quality Assurance.* Rockville, IL: Food and Drug Administration.

U.S. Department of Health and Human Services. 2006. *Guidance for Industry: Investigating Out of Specification (OOS) Test Results for Pharmaceutical Production.* Rockville, IL: Food and Drug Administration.

U.S. Department of Health and Human Services. 2008. Chapter 45: Biological drug products, inspection of biological drug products (CBER). In *Compliance Program Guidance Manual.* Rockville, IL: Food and Drug Administration.

U.S. Department of Health and Human Services. 2008. *Guidance for Industry: Process Validation: General Principles and Practices.* Rockville, IL: Food and Drug Administration.

U.S. Environmental Protection Agency. Combustion. http://www.epa.gov/epaoswer/hazwaste/combust.htm (links to Solid Waste Combustion/Incineration, and Hazardous Waste Combustion).

U.S. Pharmacopeia (USP). 2012, Sterilization and Sterility Assurance. http://www.usp.org; www.uspnf.com/uspnf/login.

Uttamlal, M. and Walt, D.R. 1995. A fiber-optic carbon dioxide sensor for fermentation monitoring. *Biotechnology* 13 (6): 597–601.

Valax, P., Charbaut, E., Dathe, J.E., Tarrach, K., Lamproye, A., and Broly, H. 2009. Robustness of parvovirus-retentive membranes and implications for virus clearance validation requirements. *BioProcess Int.* 7(Suppl. l): 56–62.

Valentine, P. 2009. Implementation of a single-use stirred bioreactor at pilot and GMP manufacturing scale for mammalian cell culture. ESACT 2009 Meeting, Dublin, Ireland.

Van den Vlekkert, H.H., de Rooij, N.F., van den Berg, A., and Grisel A. 1990. Multi-ion sensing system based on glass-encapsulated pH-ISFETs and a pseudo-REFET. *Sens. Actuators B Chem.* 1 (1–6): 395–400.

Van Tienhoven, E.A.E., Korbee, D., Schipper, L., Verharen, H.W., and De Jong, W.H. 2006. In vitro and in vivo (cyto) toxicity assays using PVC and LDPE as model materials. *J. Biomed. Mater. Res. A* 78: 175–182.

Verjans, B., Thilly, J., Hennig, H., and Vandecasserie, C. 2007. Qualification results of a new system for rapid transfer of sterile liquid through a containment wall. *Pharm. Technol.* 31: 184–195.

Vogel, J. 2005. Fast capture of biopharmaceuticals from continuous cell culture. Comprehensively Chromatography Conference, Emeryville, CA.

Vogt, R. and Paust, T. 2008. Disposable factory of tailor made integration of single-use systems. *BioProcess Int.* 7 (Suppl. 1): 72–77.

Walsh, G. 2007. Engineering biopharmaceuticals. *BioPharm. Int.* 20: 64–68.

Walter, J.K., Nothelfer, F., and Werz, W. 1998.Validation of viral safety for pharmaceutical proteins. In G. Subramanian (ed.), *Biodissolution and /Processing*, Vol. I. Weinheim: Wiley-VCH, 465–596.

Walter, J.K. 1998. Strategies and considerations for advanced economy in downstream processing of biopharmaceutical proteins. In G. Subramanian (ed.), *Bioseparation and Bioprocessing. Processing, Quality and Characterization, Economics, Safety and Hygiene.* Weinheim: Wiley VCH, 447–460.

Wang, E. 2006. Cryopreservation, storage and transportation of biological process intermediates. *BioProcess International Industry Yearbook* 2006: 78–79.

Wang, J., Moult, A., Chao, S.F., Remington, K., Treckmann, R., Emperor, K., Pifat, D., and Hotta, J. 2004. Virus inactivation and protein recovery in a novel ultraviolet C reactor. *Vox Song.* 86: 230–238.

Watler, P.K. 2009. Solving 21st Century Manufacturing Challenges with Single-Use Technologies. Presentation given at the Bio-Process Systems Alliance Meeting, La Jolla, CA; http://www.BPSalliance.org.

Weathers, P.J., Towler, M.J., and Xu, J. 2010. Bench to batch: Advances in plant cell culture for producing useful products. *Appl. Microbiol. Biotechnol.* 85: 1339–1351.

Weber, W., Weber, E., Geisse, S., and Memmert, K. 2002. Optimisation of protein expression and establishment of the wave bioreactor for baculovirus/insect cell culture. *Cytotechnology* 38: 77–85.

Wehnert, G., Anders, K.D., Bittner, C., Kammeyer, R., Hubner, U., Niellsen, J., and Scheper, T. 1989. Combined fluorescence scattered-light detector and its use in process monitoring in biotechnology. 1989 Annual Meeting of Process Engineers, Berlin, Fed. Rep. Ger.

Wei, J., Yang, H., Sun, H., Lin, Z., and Xia, S. 2006. A full CMOS integration including ISFET microsensors and interface circuit for biochemical applications. *Rare Met Mater. Eng.* 35 (3): 443–446.

Weidner, J. and Jimenez, F. 2008. Scale-up case study for long term storage of a process intermediate in bags. *Am. Pharm. Rev.* Available: http://www.american pharmaceuticalreview.com/ViewArticle.aspx?ContentID=3486.

Weigl, B.H. and Wolfbeis, O.S. 1995. Sensitivity studies on optical carbon dioxide sensors based on ion pairing. *Sens. Actuators B Chem.* 28 (2): 151–156.

Weitzmann, K.H. 1997. The use of model solvents for evaluating extractables from filters used to process pharmaceutical products. *Pharm. Technol.* 21: 73–79.

Wells, B. et al. 2007. Guide to disposal of single-use bioprocess systems. *BioProcess Int.* 6(5): S24–S27.

Wells, B. 2007. Guide to disposal of single-use bioprocess systems. *BioProcess Int.* 11: 22–28.

Wendt, D. 2003. BioTrends: Disposable processing systems: How suppliers are meeting today's biotech challenges form fluid handling to filtration. *BioPharm, Int.* 15: 18–22.

Werner, S. and Nägeli, M. 2007. Good vibrations. *BioTechnology* 3: 22–24.

WHO Expert Committee on Specifications for Pharmaceutical Preparations. 2003. Good manufacturing practices for pharmaceutical products: Main principles, Technical Report Series No. 908, Annex 4. Geneva, Switzerland: World Health Organization.

Williamson, C., Fitzgerald, R., and Shukla, AA. 2009. Strategies for implementing a BPC in commercial biologics manufacturing. *BioProcess Int.* 7 (10): 24–33.

Wilson, J.S. 2006. A fully disposable monoclonal antibody manufacturing train. *BioProcess Int.* 4 (Suppl. 4): 34–36.

Wolfbeis, O.S. 2005. Materials for fluorescence-based optical chemical sensors. *J. Mater. Chem.* 15 (27–28): 2657–2669.

Wong, R. 2004. Disposable assemblies in biopharmaceutical production: Design, implementation and troubleshooting. *BioProcess Int.* Suppl. 4 (9): 36–38.

Wu, Y., Ahmed, A., Waghmare, R., Genest, P., Issacson, S., Krishnan, M., and Kahn, D.W. 2008. Validation of adventitious virus removal by virus filtration: A novel procedure for monoclonal antibody of process. *BioProcess Int* 5: 54–59.

Wurm, F.M.. 2007. Novel technologies for rapid and low cost provisioning of antibodies and process details in mammalian cell culture-based biomanufacturing. BioProduction 2007, Berlin, Germany.

Xu, J., Zhao, W., Luo, X., and Chen, H. 2005. A sensitive biosensor for lactate based on layer-by-layer assembling MnO_2 nanoparticles and lactate oxidase on ion-sensitive field-effect transistors. *Chem. Commun.* (6): 792–794.

Yates, D.E., Levine, S., and Healy, T.W. 1974. Site-binding model of the electrical double layer at the oxide/water interface. *J. Chem. Soc. Faraday Trans.* 170: 1807–1818.

YSI Life Sciences. 2010. YSI STAT 2300, YSI 2700 SELECT. Available: http://www.ysilifesciences.com.

Zambaux, J.P. 2007. How synergy answers the biotech industry needs. BioProduction 2007, Berlin, Germany.

Zambeaux, J.P., Vanhamel, S., Bosco, F., and Castillo, J. 2007. Disposable bioreactor. Patent EP 1961606A2.

Zandbergen, J.E. and Monge, M. 2006. Disposable technologies for aseptic filling. *BioProcess Int.* 6: 48–51.

Zeta. 2009. FreezeContainer®. Available: http://www.zeta.com/DE/Produkte/Freeze-Thaw-Systeme/.

Zhang, H., Williams-Dalson, W., Keshavarz-Moore, E., and Shamlou, P. 2005. Computational-fluid-dynamics (CFD) analysis of mixing and gas–liquid mass transfer in shake flasks. *Biotechnol. Appl. Biochem.* 41: 1–8.

Zhang, R., Bouamama, T., Tabur, P., Zapata, G., Gottschalk, U., Mora, J., and Reif, O. 2004. Viral clearance feasibility study with Sartobind Q membrane adsorber for humanly antibody purification. IBC 3rd European Event BioProduction 2004, Munich, Germany.

Zhang, X., Bürki, C.A., Stettler, M., De Sanctis, D., Perrone, M., Discacciati, M., Parolini, N., DeJesus, M., Hacker, D.L., Quarteroni, A., and Wurm, F.M.. 2009. Efficient oxygen transfers by surface aeration in shaken cylindrical containers for mammalian cell cultivation at volumetric scales up to 1000 L. *Biochem. Eng. J.* 45: 41–47.

Zhang, X., Stettler, M., De Sanctis, D., Perrone, M., Parolini, N., Discacciati, M., De Jesus, M., Hacker, D., Quarteroni, A., and Wurm, F. 2009. Use of orbital shaken disposable bioreactors for mammalian cell cultures from the mL scale to the 1,000 L scale. In Eibl, D. and Eibl, R. (eds.), *Disposable Bioreactors, Series: Advances in Biochemical Engineering/Biotechnology*, Vol. 115. Berlin; Heidelberg: Springer, 33–53.

Zheng, R. 2010. The game changer, *BioProcess Int.* 8(4): S4–S9.

Zhou, J. and Dehghani, H. 2007. Development of viral clearance strategies for large-scale monoclonal antibody production. In E.S. Langer (ed.), *Advances in Large Scale Biomanufacturing and Scale-Up Production*, 2nd ed. Rockville, MD: ASM.

Zhou, J. and Tressel, T. 2006. Basic concepts in Q membrane chromatography for generous scale antibody production. *Biotechnol. Prog.* 22: 341–349.

Zhou, J.X., Solamo, F., Hong, T., Shearer, M., and Tressel, T. 2008. Viral clearance using disposable systems in monoclonal antibody commercial downstream processing. *Biotechnol. Bioeng.* 100: 488–496.

Zhou, J.X., Tressel, T., Gottschalk, U., Solamo, F., Pastor, A., Dermawan, S., Hong, T., Reif, O.W., Mora, J., Hutchinson, F., and Murphy, M. 2006. New Q membrane scale-down model for process-scale antibody purification. *J. Chromatogr. A* I 134: 66–73.

Zhou, J.X., Tressel, T., and Guhan, S. 2007. Disposable chromatography: Single-use membrane chromatography as a polishing option during antibody purification is gaining momentum. *BioPharm. Int.* 1 (Suppl. 1): 26–35.

Zhou, J.X., Tressel, T., Yang, X., and Seewoester, T. 2008. Implementation of advanced technologies in commercial monoclonal antibody production. *Biotechnology.* 3: 185–1200.

Zhou, J.X. and Tressel, T. 2006. Basic concepts in Q membrane chromatography for generous scale antibody production. *Biotechnol. Prog.* 22: 341–349.

Zhou, J.X., Solamo, F., Hong, T., Shearer, M., and Tressel, T. 2008. Viral clearance using disposable systems in monoclonal antibody commercial downstream processing. *Biotechnol. Bioeng.* 100: 488.

Ziv, M, Ronen, G, and Raviv M. 1998. Proliferation of meristematic clusters in disposable pre-sterilized plastic biocontainers for the large-scale propagation of plants. *In Vitro Cell Dev. Biol. Plant* 34: 152–158.

Ziv, M. 1999. Organogenic plant regeneration in bioreactors. In Altmann, A., Ziv, M., and Izhar, S (eds.), *Plant Biotechnology and In Vitro Biology in the 21st Century.* Dordrecht, The Netherlands: Kluwer, 673–676.

Ziv, M. 2000. Bioreactor technology for plant micropropagation. *Hort. Rev.* 24: 1–30.

Ziv, M. 2005. Simple bioreactors for mass propagation of plants. *Plant Cell Tissue Organ. Cult.* 81: 277–285.

Zlokarnik, M. 2001. *Stirring-Theory and Practice.* Weinheim: Wiley-VCH Verlag GmbH & Co. KGaA.

Zlokarnik, M. 2006. *Scale-Up in Chemical Engineering.* Weinheim: Wiley-VCH Verlag GmbH & Co. KGaA.

Zweifel, H. 2001. *Plastic Additives Handbook,* 5th edition. Munich, Germany: Hanser Publishers.

Index